SYSTEMS ARCHITECTING OF ORGANIZATIONS

Why Eagles Can't Swim

SYSTEMS
ARCHITECTING
OF
ORGANIZATIONS
Why Eagles Can't Swim

EBERHARDT RECHTIN

CRC Press
Taylor & Francis Group
Boca Raton London New York

CRC Press is an imprint of the
Taylor & Francis Group, an **informa** business

CRC Press
Taylor & Francis Group
6000 Broken Sound Parkway NW, Suite 300
Boca Raton, FL 33487-2742

© 2000 by Taylor & Francis Group, LLC
CRC Press is an imprint of Taylor & Francis Group, an Informa business

No claim to original U.S. Government works

ISBN 13: 978-0-8493-8140-9 (hbk)

Visit the Taylor & Francis Web site at
http://www.taylorandfrancis.com

and the CRC Press Web site at
http://www.crcpress.com

Library of Congress Card Number 99-26157

Library of Congress Cataloging-in-Publication Data

Rechtin, Eberhardt.
    Systems architecting of organizaitons : why eagles can't swim / Eberhardt Rechtin.
        p. cm. — (Systems engineering series)
    Includes bibliographical references and index.
    ISBN 0-8493-8140-1 (alk. paper)
        1. Organizational effectiveness. 2. Systems engineering. I. Title. II. Series.
HD58.9.R43 1999
658.4—dc21   99-26157
                                                    CIP

Preface/Introduction

Why eagles can't swim

*The eagle: a wondrous and admired creature,
a soaring master of flight.
A metaphor for excellence, strength, courage, and pride.*

This is a book about excellent organizations, large and small; eagles in their own domains, yet unsure just how to respond to global changes going on around them. They know that being excellent in one field doesn't guarantee success in another. They know that eagles can't swim, regardless of motivation. They are magnificently built for masterful flight in the air, not for catching fish deep below the water. More helpful and specific is the insight that:

> *Given an excellent organization successful in its own field
> with objectives, skills, and policies designed for that success,
> there are some things it can not do — or at least not do well.*

In point of fact, strengths in one field may well turn out to be weaknesses in another, which begs the following questions:

1. What *specific* organizational strengths lead to what *specific* things that cannot be done well by *whom* — where, when and why?
2. What, if anything, might or should be done to change that outcome if desired?

The four premises

This book addresses these questions based on four premises:

1. Organizations are complex systems, people-based, but nonetheless *systems.**
2. Every system and organization has an architecture, or "structure" broadly defined which largely determines what the system can and can't do.

---

* Briefly, elements working together to a larger purpose.

3. Systems architecting can be as applicable to the structural problems of organizations as it is to the problems of the hardware and software products the organizations create and support.
4. Systems architectural insights and techniques,* heuristics and metaphors in particular, can be effectively used to sustain the excellence of organizations, their people, and their product lines — especially during times of global competition and unavoidable change.

For better or worse, the most difficult questions facing organizations today do not have scientifically or mathematically provable solutions. Because the answers that do exist depend upon time and circumstance, they are seldom replicable. That means that the validity of the answers depends upon the experience and biases of their authors, their close associates, and, of course, on those who are asking the questions.

Origins of this book

As to this last point, this author has had the rare privilege of working as an executive in six truly excellent organizations — CalTech's Jet Propulsion Laboratory, the Department of Defense's Advance Research Projects Agency, the Office of the Secretary of Defense, the North Atlantic Treaty Organization (NATO), the Hewlett-Packard Company, The Aerospace Corporation, and as a professor in systems architecting at the University of Southern California. He has worked closely with many others including MIT's Lincoln Laboratory, the MIT Department of Aeronautical and Astronautical Engineering, the Bell System, the intelligence community, the School of Engineering of the University of Southern California, Hughes Aircraft, and TRW. Truly eagles, all.

The book was written because highly respected organizations like these — and the skilled professionals on which they depend — are now confronted by a very difficult dilemma. They must somehow maintain their vaunted excellence, accommodate the new world of global communications, transportation, economics, and multinational security, and still survive against stiff competition already in place. As they are finding out, depending upon the circumstances, the demands of excellence on the one hand and of change on the other can be cruelly irreconcilable.

Many books and case studies have been written describing business successes and failures, using such companies such as Westinghouse and GE, IBM and Microsoft, Digital Equipment Corporation and Apple Computer as detailed examples. The approach of this book differs in two ways.

First, the approach is architectural and heuristic.[1] That is, the book identifies factors inherent in an organization's structure and purposes which can

---

* Insights are perceptive understandings of the underlying nature of things. Heuristics, as used here, are the specialized insights of systems architecting — generalized lessons learned from experience made credible by examples and stated criteria. Their purpose is to *guide*. Metaphors, a related kind of insight, are a technique to *educate* through similarity.

lead to excellence in one field but to potential weaknesses in another. It then provides guidelines and insights (or "heuristics" in the language of systems architects) to address them. From determining such factors comes a better understanding of what is possible, what is changeable, and what may have to be accepted as inevitable.

Second, this book is about maintaining success in a dynamic world, not about achieving it in a static one. Most successful organizations and professionals are aware of the things they can do and the things they cannot, as long as the world around them remains much the same and as long as they don't have to reinvent themselves. Yet, few are clear on what might be best to do — or not to do — should truly major change be in the wind. For example, few executives know what to acquire or what to divest other than for the most general of financial reasons. They do know that they may be badly burned in the details of the transactions, or they may erode the human foundations of the organization's excellence. The record on massive financial mergers or divestitures is at best mixed. But they also know that over the long run, they must change and the greater the change the greater the risk to their survival.

Consequently, the new global economy, the end of the Cold War, the extraordinarily high rates of technological and medical change — and the management uncertainties of how to accommodate them — have placed truly excellent organizations in considerable jeopardy. Highly skilled professionals have seen their careers, their investment in skills and education, and their well-being subjected to downsizing, mergers, reinvention, obsolesence and, yet, at the same time, have seen unimagined opportunities in new ventures.

Government agencies have seen their purposes seriously scaled back, their functions increasingly out-sourced, their reputations unfairly maligned. Welfare, social security, and health services are being drastically restructured financially, politically, and socially. Excellent small businesses have witnessed their local and regional markets become subsumed into the maw of a global economy and the competitors that roam relentlessly in it.

And yet, computers, software, networks, medicines, DNA applications, and much better understanding of the human brain have opened a whole new world for everyone. The difficulty with the new world is that it is new — and different. In a new world, excellence also is different.

Understanding excellence for what it is

The first step in responding to the dilemma (of excellence in one context leading to weaknesses in another) is to understand excellence and its sources. It is a truism that in a chosen field and given the right conditions, excellence produces success. But, change the field or encounter a new situation and all bets are off. Excellence, in other words, depends as much on context as on inherent capability; a statement as true for individuals as for organizations.

A notably innovative, excellent company, Hewlett Packard, proved unable in the 1970s to build profitable, large-scale, complex systems. The difficulty could be traced to divisional autonomy; a strength in individual product lines, but a weakness in system integration.

A highly successful manufacturer, General Motors, failed in the 1980s to introduce new technology into the workplace. The cause was an unwillingness to restructure manufacturing line management from its original and successful outlines. General Motors was not alone; 50 to 70% of the attempts to introduce new technology into the work place fail. [MA 88]*

A not-for-profit spinoff of RAND, called SDC, tried to become a profit-maker and was instead swallowed up by a for-profit competitor. In brief, SDC had never been structured to make a profit and didn't know how. Another not-for-profit, The Aerospace Corporation, barely escaped with the trust and confidence of its major client still intact for even considering a for-profit conversion. Though being an overly restricted not-for-profit had been exceptionally difficult for the company, all things considered, becoming a for-profit, would not have been better, only different.

Could such results have been predicted ahead of time? Probably. Indeed, some were and for the most basic and understandable of reasons — the inherent conflict between present sources of success and the needs of future programs on "the other side" of the oncoming change.

The approach: recognizing and architecting organizations as complex systems

The next step toward resolving the excellence–evolution dilemma is to recognize that an organizations is a *system*; that is, it is a structure created to satisfy client purposes otherwise unachievable separately by its elements. The added client value is the justification for the system, the reason it is built.

Creating and building complex systems to add value is the province of systems architecting and, according to the fourth premise of this book, could be the province of creating and sustaining the added value of organizations, products, and people.

Systems architecting, for reasons developed in Chapter 1, is principally an art, though it does rely in important ways on the quantitative science of systems engineering. The reason for the emphasis of this book on the art aspect is that organizational issues are so complex and replicable measurements necessary for their analysis so sparse that quantitative techniques provide little practical help. For example, drawing a reasonably complete description of a real organization soon leads to an almost incomprehensible confusion of boxes and lines. Yet, somehow, the organization works. It works through the use of common, or more accurately, *contextual* sense — insights gained from experience in a particular context.[2] Almost every manager uses

---

* Appendix C lists and annotates references cited in the text in the following format: [AB 97 132], where AB are the first letters of the author's last name, 97 is the year of publication, and 132 is the page number when needed. Referenced conversations or unavailable documentation are shown as (Name, Date).

these "obvious" lessons learned individually from personal experiences in individual circumstances. Perhaps surprising to some, these individually experienced insights can be generalized and refined into the tools and techniques of an art — one of guidelines, abstractions, and analogies, or, respectively, heuristics, models, and metaphors. They also happen to be the tools of artists, philosophers, writers, doctors, lawyers, and of their crafts. Never perfect, never "provable," never optimum, they nonetheless "satisfy". They solve real problems in a real world.

The use of insights and heuristics throughout this text

Thousands of years ago, the Chinese recognized the value of insights in their familiar:

A picture is worth a thousand words.

The same value of abstracted wisdom is seen in C. W. Sooter's 1993 update:

An insight is worth a thousand analyses [SO 93]

Many readers can validate this update from examples in their own past. Indeed, they may volunteer to do so with real vigor.

Validation is a serious problem for these nonmathematical insights. The author, therefore, created a list of criteria which credible heuristics — as opposed to arguable assertions, speculations, and personal opinions — would have to meet. They are

- An insight must make sense in its original domain or context.
- The general sense, if not the specific words, of an insight should apply beyond the original context.
- An insight should be easily rationalized in a few minutes or on less than a page.
- The opposite statement of an insight should be foolish, clearly not "common sense". That is, the opposite should not be equally sensible, depending only on the circumstances.
- The lesson of an insight, though not necessarily its most recent formulation, should have stood the test of time and earned a broad consensus.

Heuristics in the systems architecting field have been subjected to these criteria for almost a decade, published in two texts, and taught to over 200 systems-experienced students, managers, and executives during that time. They have indeed stood the tests of time, circumstance, and graduate students.

Though it is only speculation at this time, it is probably true that heuristics can be even more useful if several can be used at the same time for the same problem. It also is probably true that a proposed action or decision is stronger if it is consistent with several heuristics rather than only one. And it would certainly seem desirable that a heuristic be "refineable" or particularized into a design rule or a decision algorithm for different domains.

To distinguish insights from the body of the text, they have been *italicized*, inset from the rest of text itself, and referenced whenever possible. Three have already appeared in this Preface. Many of the upcoming (200 or so) heuristics have been published, with examples, in earlier works [RE 91 and RE 97]. Many more were gathered during the years 1988 to 1994 — and are still being gathered — by graduate engineering students at the University of Southern California (USC) in the ongoing courses in systems architecting under Associate Dean Dr. Elliott Axelband. As engineers and mid-level managers, these working students had experience with, or were working full-time on, complex systems underway in aircraft, aerospace, and commercial corporations in Southern California. Their heuristics have been screened by the author to be properly documented, hard-won lessons learned by and for systems architects, engineers, and managers. Like all consensus understandings, the original source(s) are difficult to discover. But, whether by discoverer or originator, there is agreement that these are practical responses to recurring themes. They should be used with care and tailored for appropriate circumstances. As two of Susan Ruth's heuristics put it,

> *Experience is the hardest kind of teacher.*
> *It gives you the test first and the lesson afterward.*
> and
> *Experience is knowing a lot of things you shouldn't do.*
> [RU 93]

The phenomenon of quick recognition by students of insights often buried in mid-paragraph of papers or industry reports, led to the realization that the central purpose of teaching system architecting was, and is, to teach students to think insightfully. Truly, as Sooter recognized,

> *An insight is worth a thousand analyses.*
> or, more pointedly
> *Architectural insights are worth far more than*
> *ill-structured engineering analyses*
> *in synthesizing and describing inherent characteristics of systems.*

Insights serve another important function in this book. They can act as highlights of whole sections of text and as bookmarks for fast search and recovery of important subjects. Consistent with this purpose, Appendix B lists more than 100 organizational insights discussed in the text by category and page, with several of the most valued ones appearing in more than one category. Most appear in three or more places, as the same lesson is learned in a new domain or context.

Metaphors and their use

A second aid to understanding and raising pertinent questions about a complex subject or organization is the metaphor. A metaphor is a literary technique for making difficult ideas in one field more understandable and actionable by comparing them with similar ideas in a field that is more familiar. One such is given in the subtitle of this book, *Why Eagles Can't Swim*,[3] which captures the book's complex theme in an easily remembered image. It converts thousands of words about excellent organizations undergoing change into an image of a proud and powerful bird being forced by events to catch fish in streams where fish are becoming scarce.

There are other metaphors thoughout the book, notably the "desktop" metaphor for personal computers and the hardware store for a heuristics list. They, too, raise serious questions of survival. Questions such as: should personal computers contain functions not normally associated with desk tops? What are suitable metaphors for the Internet? What are the appropriate metaphors for describing the competitive world — the South African jungle? English cricket? A level playing field? Or should it be the arena in the famous poem, "The Victor's Crown" by Theodore Roosevelt in which "Credit belongs to the man who is actually in the arena, whose face is marred by dust, sweat, and blood, who strives valiantly…"

Are such questions serious? They are serious enough to be the mottos of some of the most powerful organizations and people in the world. And they lead to very different management styles, options, ethics, strengths, and vulnerabilities. The eagle as a metaphor for this book was not an idle, random choice.

The design and use of this book

Writing about complex systems and organizations in the linear, word-follows-word format of a book is a complex, ill-structured problem in itself. Different readers will read the book for different purposes at different times with different perspectives. The book literally had to be architected for and from many different perspectives.

Readers who scan before reading may check the preface, table of contents, introductions, and summaries. For other readers, theory, practice, and credibility come first — the five parts of the full text are written with that perspective in mind.

Part I, therefore, is about the basics — why management is more an art than a science and why the real test of an excellent organization is in the value it adds compared with that which it receives. In technical terms, the output at every level should be of greater value than just the sum of the input values.

Part II is about the constraints imposed on and the opportunities offered to an organization by the outside world. This subject might seem a diversion from the theme of this book (what a compay can do *for itself*) except for the fact that there is little point in internal optimization if the outside world is likely to preclude it. The message of Part II is all too easily forgotten as excellent organizations, eager to capitalize on their strengths, charge ahead into sectors that they do not understand, for which they are unprepared, and where their own strengths all too easily become weaknesses.

Part III is a discussion of *internal* factors that can either extend or diminish new opportunities. The focus, as said before, is architectural not personal, as important as the latter may be.

Part IV brings the insights of systems architecting to bear on the questions posed at the beginning of this Preface and Introduction. It explicitly illustrates how systems architecting insights can aid in reorganization. These first four parts lay the foundation of what is called here "the art of organizational architecting."

Part V, the payoff, addresses the central question that excellent organizations face in an era of global change — why, when, what, and how should it either stay the present course or change to a new one. Clearly, excellent organizations wish to do more than just survive, they want to remain among the excellent.

Chapter 10 addresses the why and the when in the context of an evolutionary world, one typified by the 40 years of the Cold War between 1945 and 1985. It concludes with an example of architecting a reorganization. Chapter 11 addresses the what and the how in the context of a more turbulent period, one typified by the years after 1985 and into the twenty-first century. This final chapter takes on the more difficult architecting problems of downsizing, startups, mergers, acquisitions, divestitures, and retention of professionals.

Appendices A, B, and C present still more perspectives. For those that prefer credible propositions, backed up by examples and commentary, see Appendix A. Those who want to create their own tool kits of insights, see Appendix B, which lists insights in the text organized by architecting subject. Still other readers, often an academic group, look to the citations of other authors and works that have been used as foundations or starting points. Appendix C, therefore, lists and annotates references cited in the text, shown both there and in the text in the following format: [AB 97 132], where AB are the first letters of the author's last name, 97 is the year of publication and 132 is the page number when needed. Referenced conversations or unavailable documentation in text are shown as (Name, Date). Following the appendices are a conventional glossary and indices by author and subject.

All these perspectives are interlinked, none are quantitative, and none are either right or wrong. All suggest approaches but none are mandates. And just like architecting itself, they take a user's wisdom to bring them to life.

The central purpose of the book is built into the book's structure. However, other purposes are also served. Arguably, the most important additional purpose is to encourage readers to view their organizations in a number of different ways. Chapter 1 views organizations as complex systems; Chapter 2, as creators of emergent values; Chapter 3, as competitors; Chapter 4, as partners with government; Chapter 5, as sets of beliefs; Chapter 6, as structures; and Chapter 7, as a set of interlocking decisions. Each is important depending upon the circumstances, the problems to be solved, and the background of the viewer. It is hoped that readers will use them to gain new perspectives when reviewing their own organizations. They can be a useful way to see problems and opportunities not visible before, even in excellent organizations, by the old hands in the game.

No author wishes to think that any of the chapters are more important than others, of course. But for readers who like to read the last chapters of a novel first and then go back to the beginning to enjoy the total, the "back of the book" starts with Chapter 9.

This architecture for a book necessarily contains redundant elements (yes, the reader will see the same insight more than once) but in different contexts and applications. Like strands woven into a rope that keep reappearing along its length, these interwoven elements, because they are tightly bound to each other, make the architectural rope stronger than a set of separated strands.

The intended readership for this book

This book is primarily written for professionals and managers who are in excellent — or potentially excellent — organizations faced with the prospect of unexpected change. Perhaps the market is expanding or contracting far faster than expected. Perhaps new regulations and laws are changing the name of the game. Perhaps a mature technology is being supplanted by a dramatically different one. Perhaps the chief executive is retiring and a new one, with different ideas, is coming onboard. Perhaps a merger is in the wind. A few decades ago, before 1985 of course, a half-serious, half-joking question making the rounds in the Defense Department was, "What happens if peace breaks out?" So much for evolutionary stability....

Clearly in today's world knowing what is practical and what is not practical, and why each may be so, can be critical both to sustaining the success of the company and to the careers of its members.

The book also is designed to be used as a text for business administration and engineering students, for strategic planners, for entrepreneurial investors who rely on heuristic fundamentals in decision making, and for legislators and judges who want to minimize the effects of the "law of unexpected consequences" on their actions.

For those readers in companies for which excellence is at best only a hope, this book may make the road ahead seem still harder. Not only must the barriers to excellence be hurdled, there are others to be faced once excellence is reached.

Acknowledgments

In a very real way, this book is the result of the technical management experiences of seven, now-retired, top executives, representing 15 major organizations: Dr. Max T. Weiss (of the Bell Laboratories, TRW, and SVP of The Aerospace Corporation, Northrop, and Northrop-Grumman); Dr. Albert D. (Bud) Wheelon (of TRW, the CIA, and President, Hughes Aircraft); Allan Boardman (of The Aerospace Corporation and who conceived the "eagle" metaphor); Robert L. Cattoi (of Collins Radio and Rockwell-Collins); Dr. Elliott Axelband (of Hughes Aircraft and USC); Dr. Robert Spinrad (of Brookhaven National Laboratories and Xerox Corporation); and myself.

In addition, key insights in the text can be attributed to many other associates, fellow executives, and friends over the years: Arthur Raymond (Douglas Aircraft), Harry Hillaker (General Dynamics), Dr. Brenda Forman (Lockheed Aircraft), Ben Bauermeister (Hewlett-Packard and CEO, Elsewhere, Inc.), Harry Bernstein (AT&T), Bob Moot (Department of Defense), Bill Drake (The Aerospace Corporation), my son Mark E. Rechtin (*Automotive News*), and Dr. Mark W. Maier (Hughes Aircraft and University of Alabama/Huntsville). Another and long unpaid debt is owed to those mentors and executives who provided the wisdom and patience I needed as I learned, used, and attempted to articulate the art of systems and organizational architecting. Over 200 students contributed greatly to the field, teaching this author at the same time he taught them, and producing remarkable and original reports, many of which are cited in Appendix C. May they and others of their generation not have to say as I and others of our generation have had to say,

I wish we had known this before we started...

Now that all this has been said, what is written herein is solely my responsibility. After all, I didn't always agree with my friends and associates nor they with me.

Eberhardt Rechtin

Notes

1. Heuristic: Concerned with insights and guidelines as opposed to surveys, measurables, and closed-form solutions. Heuristic is derived from a Greek word meaning "guide" and is used here in the sense of providing a lesson learned from experience for use in present situations. For heuristics as organizational tools, see Chapter 8.
2. In more recent fields of engineering, these insights often take the form of "lessons learned" and carry with them the experience and credibility of the authors. Some of the best supported insights in the commercial aircraft field, for example, are the 20 given by John E. Steiner of which seven are readily

generalizable and are echoed in the text and Citations list of Appendix C. [See STE 83 32-3] They concern defining success, red teams, a broad spectrum of requirements, compromise, and what this text calls the story of the six eagles. The other 13 lessons are more specific to the aircraft domain. Another less formal list was generated by Larry Bernstein, the retired chief technical officer of AT&T Network Systems, in a fascinating collection of one-liners from Mark Twain to Dwight Eisenhower, and of more personal insights on software development. [BER 97] Several are referenced in the Insights List in Appendix B.

3. See Part Four introduction.

The Author

Eberhardt Rechtin, Ph.D., recently retired from the University of Southern California (USC) as a professor with joint appointments in the departments of Electrical Engineering/Systems, Industrial and Systems Engineering, and Aerospace Engineering. His technical specialty is systems architecture in those fields. He had previously retired (1987) as President-emeritus of The Aerospace Corporation, an architect–engineering firm specializing in space systems for the U.S. Government.

Rechtin received his B.S. and Ph.D. degrees from CalTech in 1946 and 1950, respectively. He joined CalTech's Jet Propulsion Laboratory in 1948 as an engineer, leaving in 1967. At JPL he was the chief architect and director of the NASA/JPL Deep Space Network. In 1967 he joined the Office of the Secretary of Defense as the Director of the Advance Research Projects Agency and later as Assistant Secretary of Defense for Telecommunications. He left the Defense Department in 1973 to become chief engineer for Hewlett-Packard. He was elected as the president and CEO of aerospace in 1977, retiring in 1987, and joining USC to establish its graduate program in systems architecture.

Rechtin is a member of the National Academy of Engineering and a Fellow of the Institute of Electrical and Electronic Engineers, the American Institute of Aeronautics and Astronautics, the American Association for Advancement of Science and is an academician of the International Academy of Astronautics. He has been further honored by the IEEE with its Alexander Graham Bell Award, by the Department of Defense with its Distinguished Public Service Award, NASA with its Medal for Exceptional Scientific Achievement, the AIAA with its von Karman Lectureship, CalTech with its Distinguished Alumni Award, and by the (Japanese) NEC with its C&C Prize. Rechtin also is the author of *Systems Architecting, Creating and Building Complex Systems*, Prentice-Hall, 1991, and *The Art of Systems Architecting*, CRC Press LLC, 1997.

Dedication

to

Max T. Weiss, who asked for and inspired this book,

to my students from whom I learned so much,

and

to my partner, my astute editor, and beloved wife,

Deedee Denebrink Rechtin, who with me has lived

the 50 years of experiences reflected herein

Contents

Chapter 2

Part II

Chapter 3

Chapter 4

Part III

Chapter 5

Chapter 6

Chapter 7

Part IV

Chapter 8

Chapter 9

Part V

Chapter 10

Chapter 11

Appendices

part I

---

*Treating organizations
as systems*

Key terms and concepts in systems architecting

Of course we know the meanings of those terms.
It is just that no two of us agree on what they are!

Introduction

Different words mean different things to different people. The key words used in systems architecting are no exception. Excellence, success, and value can easily be confused. However, they are quite different. Systems, structures, and architectures are sometimes interchanged. The definitions of architecting, designing, and engineering have been the subjects of debate for at least 200, if not 2000, years. Complexity, nonlinearity, and chaos are much more closely related mathematically and systemically than we used to think. And, for better or worse, most have multiple definitions in the standard dictionaries, depending upon usage and context. Few have exact synonyms.

Therefore, it is very important in accomplishing the purposes of this book that its terms be defined early on and up front. To aid in their understanding, the next section (Definitions of architecting terms) is deliberately organized very differently from the alphabetical listing in most dictionaries. Related terms are grouped so that differences between them can more easily be seen. And the definitions are tailored to their usage here. Excellence, for example, is not only defined early, it is amplified and discussed at some length in the sections that follow.

When in doubt of the meaning of these key terms, please consult the definitions in the following section or in the Glossary at the end of the book.

Definitions of architecting terms

Excellence, success, value, and risk

These four terms share a common property — all are in the eyes of the beholders, the most self-interested of which are stakeholders. None of the

3

terms are readily quantifiable, that is, characterized by replicable measurements. All depend strongly on context.

Excellence: The defining characteristic of organizations of repute, responsiveness, professionalism, and dedication.

Success: The delivery to a satisfied client of a promised set of results. Excellence and success are not necessarily the same; one of the reasons for writing this book. Success is dependent on context, including organizational form and purpose.

Value: Approximately the same as worth. In economic terms, value is what one is prepared to give up to own it. Greater than monetary alone, it includes reputation, careers, family, and friendships. In architecting terms, the most important value is that perceived by the client.

Added value: The difference between the total value at the end of a process compared with the total value at its start.

Emergent value: Value created solely by ("emerges from") the relationships between elements of a system. A value in addition to that created by the elements themselves. A value not existent in or obtainable from any of the subelements separately.

Emergent function: An emergent value expressed as a capability or service. (Example: Transportation is an emergent function of the assembled parts of an automobile.)

Risk: The perceived possibility of harm or loss. Generally refers to system- or enterprise-level losses of purpose or function.

Systems, structures, synergy, and architectures

These three terms are all function-based forms in which the value of whole is greater than the sum of the value of the parts.

System: A construct or collection of different elements that together produce results not obtainable by the elements alone. The elements, or parts, can include people, hardware, software, facilities, policies, and documents; that is, all things required to produce system-level results. The results include systems-level qualities, properties, characteristics, functions, behavior, and/or performance. The value added by the system as a whole, beyond that contributed independently by the parts, is primarily created by the relationships among the parts; that is, how they are interconnected.

Systems (plural): A term used to describe an hierarchy of elements each of which is not only a system but a supersystem to those below it and a subsystem to those above it.

System (or systems) approach: A management process in which virtually all decisions in all elements and subelements are made based upon the effects on the system and its functions as a whole.

Organization: A system in which people are the most numerous part and which is so organized as to create products and/or services of value to others. A general term including companies, firms, government agencies, universities, military services, and cultural groups.

Structure: Something put together from parts that are interconnected into a complex entity.

Structuring: The process of creating structures; organizing some order out of disorder.

Ill-structured: Disordered, unconfigured, disorderly, or unorganized, as in an "ill-structured problem." Characteristic of the situation in the early phases of architecting when the client's desires and architecting feasibility have not been reconciled.

Architecture: The structure — in terms of components, connections, and constraints — of a product, process, or element. Includes all elements of a system that determine its form and function. By definition, each system has an architecture, explicit or not, which can be viewed from many perspectives.

Architecting, engineering, and designing

The common property of these terms is that they are all processes for creating and making useful things. It should be noted that architecting can be done by other than formally trained architects, engineering by other than engineers, etc. It is the processes that are primary here, not the job description of its practitioners.

Architecting: The process of creating and building architectures, especially during the conceptual and certification phases. Generally synthesis-based, insightful, and inductive. [RE 91 vi]*

Architect: A person doing architecting. A professional who is skilled and working primarily in the art and science of architecting and occasionally in related fields such as the applied science of engineering.

Engineering: The process of applying science and mathematics to practical ends. [WE 84 433] Especially during the engineering design, integration, and production phases. Generally analysis-based, factual, logical, and deductive. [RE 91 vi]

Engineer: A person doing engineering. A professional who is skilled and working primarily in the applied science of a branch of engineering and occasionally in related fields such as architecting and design.

Designing: A difficult term to define. One dictionary [WE 84 367] gives eight different definitions from "conceiving in the mind" to "creating in a highly skilled manner." For the purposes here, designing generally means creating smaller objects in more detail with more limited

---

* Appendix C lists and annotates references cited in the text, shown both there and in the text in the following format: [AB 97 132], where AB are the first letters of the author's last name, 97 is the year of publication, and 132 is the page number when needed.

purposes, fewer interrelationships, and more artistic judgments than does architecting in a large-scale system and usually confined to one major medium at a time.

Designer: A person doing designing. A professional who is skilled and working primarily in the craft of design and, occasionally, engineering.

Complexity, nonlinearity, and chaos

These terms share the property of describing a confused turmoil often produced by fairly simple processes. Until recently all were viewed as a kind of "noise", treatable only by statistical averaging. They are now known to be partially structured and, hence, limited in their range of seemingly disordered behavior.

Complexity: The degree of intricacy of a system so interconnected as to make analysis difficult or impractical. The more interconnections, the more complex the system.

Nonlinearity: An element in which the result of introducing two inputs is not the same as the sum of the results of introducing them separately; in other words, they intermix. A necessary condition for chaotic behavior. As in, "A nonlinearity is a prerequisite of chaotic behavior."

Chaos and chaotic behavior: A relatively recent class of complex or "noisy" behavior now known to result from nonlinearities, fixed time delays, memory, and interconnections. Produces such phenomena as violently unstable network gyrations, nonreplicability of organizational behavior, a stock market volatility, and possibly business cycles.

The necessity of excellence in complex systems

Excellence is more than a generally good idea, it is an essential attribute of successful complex systems. As complexity grows, the effect of an otherwise inconsequential discrepancy or failure in one element can propagate to others, creating serious and unanticipated system level behavior that can be difficult to trace to its source. A heuristic, originally from the field of quality assurance, says it well:

> *An element "good enough" in a small system, if unchanged*
> *is unlikely to be "good enough" in a more complex one.*

A particularly timely example in this information age is the dilemma of so-called reusables in software. The rationale behind reusables is that software development is so costly and time consuming that advantage should be taken of routines (self-contained pieces) that have already been developed by reusing them in new software development and applications. The difficulty that is encountered is that the "reusables" almost always were developed with now-forgotten assumptions, criteria, and computers for products that were smaller, less complex, and less sensitive to failure than the one under consideration. Timing errors in the original routines, for example, were in a range that could be accommodated easily enough so that they were discounted as unimportant and, hence, forgotten. Later these routines were installed in a much larger, more complex, more important product with the consequence that nothing seemed to work. But one response is to avoid reusables until they have been subjected to the same tests and evaluation as all others in the new product. A further screening might require prior use in the same product *line*.

Other reusables that have had the same generic problem are auto and aircraft parts, antennas, batteries, and even solid rocket designs. The importance of this insight for excellent organizations considering changing their product lines is clear enough — the reusables are likely to cause trouble, they create unknown constraints on the new product line architecture, and, if modified in any way, they lose any warranty guarantees they might have from their suppliers.

Clearly, building satisfactory, complex systems with less than excellent processes and people is a sure and expensive way to failure. This fact of life is just as true for organizations as for communication systems, space systems, or microprocessors. A large high tech company with administrative support policies from its early beginnings in a small shop is bound to run into trouble in quality control, response to customer needs, and efficient manufacturing. [HA 88 and WO 90]

The danger in being excellent

The danger in being excellent is hubris, the arrogance of excessive pride. A very natural inclination of excellent, successful organizations (and of superstar professionals everywhere) is not only to believe that they can do anything, but that they can do it well. After all, they are on record as better than the norm. The corresponding assumption is that excellence is a thing unto itself, pretty much the same for any situation. Unhappily, this belief is an unstated assumption in the establishment of many conglomerates and mergers of previously excellent firms. Too often, it also can be the downfall of business administrators taught that an excellent business manager can manage any business, regardless of its technical content.

Yet, on the face of it, such assumptions are unsupportable. Superstar basketball players don't become chessmasters very often. Concert violinists, if they are wise, don't play contact sports. Innovative companies find it very difficult to duplicate their initial success once their product line becomes something others can duplicate, when cost becomes all important, and when perspiration displaces inspiration.

Serious attempts at defining excellence

Excellence has been assumed to be not only a necessary but presumably sufficient attribute of successful systems and organizations for a long time.

In effect, the assumption said that if a company was excellent, it would succeeed, and conversely, if it was successful, it must be excellent. The assumption became a serious subject of study in the U.S. when excellent American companies were faced with global competition from clearly successful companies elsewhere using a much different definition of excellence. It came to public attention in 1982 with the publication of *In Search of Excellence* by Thomas J. Peters and Robert H. Waterman, Jr. [PE 82] It is a survey of dozens of companies, generally regarded as excellent, from which eight attributes of excellence are derived. The eight are:

- A bias for action
- Close to the customer
- Autonomy of entrepreneurship
- Productivity through people
- Hands-on, value driven
- Stick to the knitting
- Simple form, lean staff
- Simultaneous loose–tight properties.

An additional attribute, intensity of beliefs, was added in the text. Limits of the book were acknowledged; namely, the condensation in the interest of simplification to only 20 of more than 60 companies, to companies whose sales were more than $1 billion, and to 8 of 22 attributes. A notable omission in Peters' list was the quality of the organization's *product*, raising the interesting question of whether Peters assumed that an excellent organization, as he defined it, automatically produced an excellent product.

The omission of the quality of product was remedied by J.M. Juran [JU 88], defining quality to have two parts: product performance (product satisfaction) and freedom from deficiencies (product dissatisfaction). Even with the improved definitions, Juran's prescriptions for a quality product differed to some extent from those for excellence of Peters and Waterman.

Notice the contrast between the company-focused excellence assessment of Peters and Waterman and the customer-focused success one of Juran. The importance of a customer focus was enhanced by subsequent studies of financial performance. Not all *In Search of Excellence* companies paid off well to stockholders, at least not within a year or so of publication. About half were in trouble.

One of the best lists of attributes of product quality was written much earlier, in 1951, by Arthur Raymond, chief engineer of the famed Douglas DC-3 aircraft. He proposed that an effective program contain:

1. A proper environment: conducive work environment, funding, and customer relations.
2. A good initial choice: what is needed vs. what is possible.
3. An excellence of detailed design.
4. A thorough development and debugging.

    5. A follow-through with assistance of operating personnel.
    6. A thorough exploitation.
    7. A correct succession: properly timed introduction of the new model.
    8. An adaptiveness, i.e., an ability to cope with the unexpected.

With respect to future changes, either evolutionary or radical, take particular note of qualities 7 and 8. Neither were mentioned in the previous definitions of excellence, a potential downfall if unrecognized by presently excellent organizations.

    Good as Raymond's 48-year-old list is, it must be judged incomplete in today's more complex world of tort liability, mammoth governmental projects, and changing public perceptions of what is and is not acceptable, much less excellent. On the other hand, the advent of knowledge management systems, of better controlled inventories, and of widespread access to previously unavailable information, any time, anywhere, on virtually any subject, means that many obstacles to success have been demolished — if one pays attention.

    Whatever the list, or others that the reader may prefer, three features stand out. The definitions of excellence and product quality are not alike, though some attributes are closely related, and each list contains more than one element (typically the oft used 7 ± 2). So, the question is, assuming each list is reasonably complete in its own context, how many elements can be omitted while still obtaining both excellent and successful results?

    In truth, a remarkable number of systems and organizations have indeed succeeded contrary to expectations and in spite of lacking one or more of the listed attributes. A modified heuristic from the world of testing new aircraft is suggestive of how many omissions can be tolerated and why.

> One *failure (or omission) acting alone can usually be corrected*
>   *in real time, the system (organization) will probably*
>   *survive.*
> Two *acting together can usually be corrected, but not in real time.*
> Three, *because of their nonlinear interactions, may never be*
>   *corrected. The system (or the pilot) will crash before the*
>   *cause can be determined.*

    To illustrate, and referring to Arthur Raymond's list, many successful products have been the result of designers starting over (failing criterion No. 2). The Apollo Lunar program bypassed critical subsystem tests and left no successor designs (omitting Nos. 4 and 7). However, few in the author's knowledge missed out on three of Raymond's criteria and still succeeded.

    Fortunately for this author and this book, there is a way out of the difficulty of defining excellence precisely. When all is said and done, few readers have difficulty distinguishing between excellent organizations and those that are not. So, choose what you believe to be one or more excellent organizations, look at them with the perspective of systems architecting and judge their future for yourself. Such was the procedure used in the University of Southern California's systems architecting classes to good effect in assessing products and projects familiar to the participants.

Two basic concepts: the essence of Chapters 1 and 2

Two basic concepts, important in creating systems and treating organizations as systems, will be introduced in Part I. The first concept is the unusual nature of complex systems in which analysis, scientific methodology, and the applied sciences — while unquestionably necessary — aren't sufficient, and why. The second concept is the importance of emergent functions in adding value at the systems level beyond that of its separate elements.

These concepts are essential to understanding organizations as systems because without the first, managers can engage in time-consuming searches for more information when none exists. The experienced manager is all too familiar with the idea that if a problem is not understood, then go out and get more data.

Without the second, there is little on which to judge whether a given level in an organization should exist, be increased in responsibility, or be eliminated. If it does create value, that is, perform a necessary function, how is it distributed if that level is eliminated? Do subordinates, superiors, or an enlarged staff pick it up? In any case, is the action perceived by the customer to add or subtract value from the product?

chapter one

The unusual field of complex organizations

"If it can't be measured, it isn't science."
Lord Kelvin (William Thomson) ca. 1871

Introduction

Today, when science and mathematics are so essential in so many fields, it is unusual to find a field in which science and its methodology don't rule. The field of complex organizations is one of those few. In it, the hallmarks of science — replicable measurements and clearly defined boundaries — are rare. Instead, ill-structured problems, ill-defined boundaries, indeterminate interactions, and unexpected consequences are the norm. As might be expected, such factors largely determine what excellent organizations can and cannot do. Consequently, they are the primary interest of this chapter and this book. Factors other than these are well described in the literature of management science and decision theory.

Where science is not enough and why

"There is often a willingness to leave untouched the most important issues in order to deal objectively with those that can be objectively quantified." [PO 95]

Science is not, indeed cannot be, sufficient in treating complex organizations because of the differing natures of both. Science is based on replicable measurements made under similar conditions. If original measurements or conditions cannot be replicated by someone else using similar but different equipment, the measurements and the conclusions that come from them are rejected as unproven and unconfirmed. As such they can't be used as evidence in any situation calling for scientific confirmation. This is not to say that the conclusions might not be true. They, in fact, *might* be true, but they can't be confirmed as such. Clearly, confirmation by replicable measurement is a major asset in any field, and many fields try mightly to incorporate it in the interests of acceptance and credibility.

Unfortunately, depending on one's point of view, scientific confirmation is almost impossible in the typical world of complex organizations. Trying to make replicable measurements is impractical if not impossible for most complex systems.

Complex systems and organizations, by definition and from practical experience, are just too intricate and interconnected for realistic, quantitative analyses. There are too many factors that change with time, too little time to collect the required information, and too few precedents from which data can be retrieved. At the project level, data suitable for statistical analysis may be too costly or take too long to accumulate; indeed, the immediate project may be long since over before suitable data can be collected. It would take thousands of flights of a satellite, for example, to verify a specified 10-year lifetime with a 99% probability of data reception at a confidence level of more than 90%. Data critical for informed decisions will certainly not be provided voluntarily by competitors or by unfriendly foreign governments. The Soviet Union was not about to provide the U.S. with its advanced weapons strategy to help the U.S. create an efficient defense acquisition program. And, not the least, few organizations, unconstrained, remain static long enough to be measured.

Where well-defined boundaries are the exception

It is in the nature of human endeavors to try to divide things into neat packages with well-defined boundaries around each. Nations are drawn on a map with sharp boundary lines encompassing them. Universities are divided into departments of chemistry, physics, business administration, and so on. Companies draw organization charts that partition and subdivide tasks, disciplines, and projects.[1]

Providing that the problems, conditions, and tasks are not too complex, all of these boundaries have served remarkably well. Indeed, the applied science of systems engineering depends on just such well-defined, albeit artificial, divisions and interfaces.

Artificial boundaries, as might be expected, create problems of their own. Geographical boundaries have caused conflicts from neighborhood squabbles to world wars. Discipline boundaries in universities, as between electrical engineering and chemistry, make interdisciplinary programs like electrochemistry very difficult to manage. Organization charts more often create competition than cooperation. "Turf wars," as they are called, are at the core of university, company, military service, and agency politics.

Complex systems and organizations are particulary vulnerable to how boundaries are specified because, as pointed out earlier, the interrelationships across the boundaries are what create the added value of system-level functions. In organizations, creating value equates to influence. No wonder that complex systems are inherently difficult to partition. Partitioning an organization by definition means partitioning responsibility and authority, a very sensitive subject to the many managers who would much prefer that their authority match their responsibility — even as they understand why that match is seldom practical. It is impractical because systems and organizations require elements that work together. That in turn means that no element can be absolutely autonomous in responsibility or authority, much less in both. All are affected by and affect at least one other element. The insight:

> *In complex organizations, almost everything is connected*
> *to everything else, directly or indirectly.*

Indeed, if it weren't, it would not, could not, be a member of the organization, nor be able to contribute to its objectives or product lines.

Overlapping regions of interest and responsibility

Connections come in many forms from the simplest plug-and-socket combination to large areas of overlapping responsibility, authority, and vested interests. One of the better ways of visualizing these overlaps is by the use of Venn diagrams.* Venn diagrams show which elements are involved, in which regions, and to what degree. For example, Figure 1.1 shows how the different elements of a satellite relate to each other. The more the overlap, the greater the mutual interest. The payload in the lower left of Figure 1.1, for example, is shown as interacting primarily with structures, guidance, control, and communications — but only secondarily with command, propulsion, and the launch vehicle. Guidance and control interact with all the other elements, thus, only guidance and control can occupy the center of the diagram. By implication, they also occupy the architectural, functional, and organizational center.

A similar Venn diagram for an organization, Figure 1.2, shows the degree to which company units[2] might intereact to achieve such common goals as profit, government relations, union negotiations, and public image. Compare this figure with the hierarchical organization chart of Figure 1.3. Single and even double overlaps can be seen in the Venn diagram of Figure 1.2, none in the organization chart of Figure 1.3. The Venn diagram emphasizes operational relationships in contrast with the reporting and directing emphasis of the organization chart. Clearly, both perspectives are important. Omission of either, in effect, puts it out of sight and mind of management. A worthwhile exercise for readers is to overlay Venn diagrams on their own organization charts. This exercise will quickly point out operational gaps and

---

* Venn diagram: A sketch using circles or enclosed areas positioned to show regions of influence. Overlapping areas signify overlapping responsibility, authority, and interest.

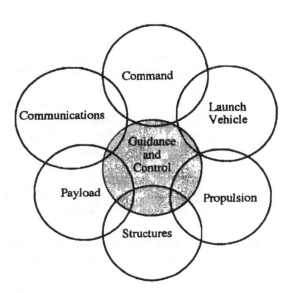

Figure 1.1 Spacecraft subsystems. (Shaded area indicates computing required, as well as indicating mutual interests.)

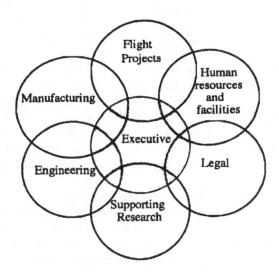

Figure 1.2 Organizational units. (Shaded area indicates computing required, as well as indicating mutual interests.)

overlaps by suggesting corrections not apparent from their organization chart.

A diagam similar to Figure 1.2 helped the CEO of The Aerospace Corporation in 1977 understand that flight projects with strong funding and supporting research with rapidly declining funding needed to have a closer connection than organization charts and Venn diagrams indicated. Some

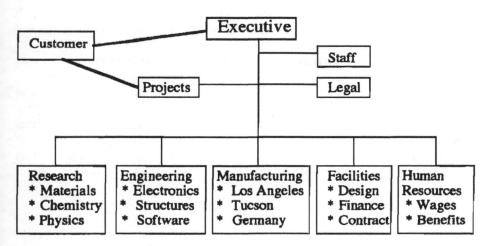

Figure 1.3   Typical hierarchical organization chart. (Shaded area indicates computing required, as well as indicating mutual interests.)

special mechanism was needed to reduce that gap. The mechanism, incidentally, was for the researchers to write the introductory paragraphs of their research reports in the "language" that projects best understood by beginning each report with the relevancy of the work to project needs. At the end of the report the researcher needed to sign off so that project people would know exactly whom to contact for further information. Importantly, in the middle of the report, the language of the researcher didn't need to be changed at all. This mechanism worked even better than expected. User friendly and much more readable research results were often expedited to the projects in one direction and research funds from the projects almost immediately increased in response. Both benefited. Even the CEO could now understand the research reports. Hardly revolutionary, but effective, nonetheless.

A recent case in technical management also can be instructive here. The story is true but it is also a good metaphor for an equivalent organizational management issue. The technical problem was to decide whether computing and management of a space satellite should be physically centralized or distributed among the propulsion, communication, instrument, structure, and guidance and control segments. The engineering conclusion was that a centralized computer was more efficient, weighed less, and was simpler to design. But procurement professionals pointed out that if all computing were centralized, subcontractors who built and supplied the subsystems would not be able to deliver tested and working devices on time and on budget. They best they could do would be to deliver untested components — literally boxes of parts. If the computing elements were distributed evenly enough to accomplish full subsystem testing prior to delivery, the chances for success would be much better. This point of view, easily visualized from the Venn diagram of Figure 1.2, prevailed. The final design showed computing appearing in both the subsystems and in a smaller central executive computer.

The final result was a clear success. The chosen configuration was much closer to being on time and on budget, though software engineers for the executive computer did have to learn much more about the software languages of other elements than ever before. Further, overall management was simpler because the interface problems among the elements were simpler.

Briefly, the managerial decision to distribute computing capability made the difference between what the project could or couldn't do well. With a few word changes, the story could just as well have described the turf battles over where computing capability should be located in an organization — in mainframes located in the computer division or in workstations located in the users' offices, or in both.

The necessity of multiple perspectives

The point to be made here will be made a number of times in this book. Organizations, as systems, can be *and should be* viewed from several different perspectives, depending on the problems to be addressed, the tools to be used, the technical and social disciplines involved, the stakeholders to be satisfied, and even the language of preference. Mark W. Maier, for example, calls for at least six different views in order to reasonably describe a system architecture, of which its form is only one.[3] As suggested before, viewing an organization as being equivalent only to its organization chart hides from view its underlying objectives, purposes, policies, and interconnections, both internal and external. It will certainly obscure what it can or can't do well. If

> *Success is in the eyes of the beholder*
> and there are many of them! [RE 91]
> and each beholder has a unique perspective,
> then there can be as many perspectives as beholders.

The law of unexpected consequences

> *Expect the unexpected* — or die watching it happen.

Managers know this heuristic as, "Beware the unknown unknowns." Most of them learn quickly to expect uncertainty and scarcity of key information in key areas. *Un*known unknowns arise in areas never imagined by either the managers or any other stakeholders. Many arise from unexpected interactions among known elements or factors, none of which alone could produce the observed consequences. Organizational examples include subcontractors going out of business, clients abrogating long-standing relationships, acquisitions of or by a competitor, or major and abrupt deletions of governmental or client funding.

Organizations experience boom and bust, go unstable or sluggish, or become chaotic[4] before they belatedly realize, if ever, that their structures and architectures have long contained all the necessary ingredients for just such behavior under just the right (or wrong) circumstances. Needless to say, when such behavior occurs in an excellent organization under the pressures of an impending radical change, the impact can be traumatic. How can an excellent organization come unglued *before* the change? The insight gained from mathematical research in an otherwise very different context was that:

> *Very simple natural phenomena can create what appears to be*
> *hopelessly complex organizational behavior and confusion.*
> It is indeed a strange world, in which
> *Constants aren't and variables don't.* (Burkett, Wm. C. 92)

Legislators, familiar with this phenomenon in their own contexts, call it "the Law of Unexpected Consequences."[5] For good reason, experienced legislators and skilled managers everywhere greatly prefer incremental change to drastic overhaul, regardless of a widely-perceived need for the latter. The hope is that if the change is not too big, then most unexpected consequences can still be reversed.

Indeed, just such an incremental approach was adopted at the end of the Cold War by the U.S. Secretary of Defense for making mandatory reductions in the defense budget, and for the same reason. In that case, controlled incrementality seems to have worked. Most likely the reasons it worked were that the necessity of the reductions was generally understood, the total reductions could be achieved over enough time that the overall situation remained stable, and, most important, the budget dynamics of the defense world were known and could be managed. High-level stability was indeed maintained.

A very different and potentially far more dangerous example from the economics field is the sudden appearance of a Federal Government surplus in 1998 when a sizeable deficit had been anticipated by the most respected agency in the government only a year before. Suddenly there was a surplus! In this case, what seems to have happened is that a series of independent factors, some of which had been predicted to produce a contrary result, produced instead the unexpected and still unexplained one, the surplus.

To economists, this result is disquieting precisely because it is unexplained and resists conventional analysis.

The point to be made here is not what to do because of the unexplained surplus but what conclusions can be drawn from its existence and why should companies, presently doing well, even care? The more important observations, at least, would include:

- The economic situation is unprecedented, complex, and demonstrably unstable.
- There are no unassailable explanations, but it is very likely that irreversible, interrelated events will be more important in determining

the outcome than present strengths and weaknesses of the economy itself.

- Lacking replicable measurements and situations, neither scientific nor economic analysis can be expected to provide credible predictions.
- For planning purposes, abrupt and radical change should be treated as certainly possible. Incremental responses will be inadequate. Unexpected consequences will be life-threatening.

Excellent companies, first of all, should care because the observations are worrisome and, second, because not many companies have made any plans to deal with them. Some suggestions for responding to these observations can be found in Chapters 10 and 11 of Part V. But, for this first chapter, the intended messages are that (1) radical, short-notice change can indeed occur unexpectedly and (2) their causes and consequences may not be understood for many years thereafter.

On managing technology

Change can occur for many reasons in many ways, some unknown until long afterward. The economic case just given is only one illustration. Another comes from the unpredictable field of research, an activity with the potential of greatly assisting before and during radical change. But managing unpredictable research would seem to be a contradiction in terms. It, most certainly, cannot be done "by the numbers." The present difficulty is that there haven't been enough long-term studies to generate even simple insights into what works or doesn't.

One of the few long-term studies on this subject was the matched pair of military technology studies, one attempting to predict the future 20 years hence, the other reviewing those predictions on their twentieth anniversary.

The first study, called Project Foresight, was done shortly after World War II. Impressed with the impact of science and prior research on the course of that war, a group of eminently qualified scientists attempted to predict, as imaginatively as possible, what the next 2 decades might produce in new defense systems. It was based on the presumption, later codified into budget regulations, that basic research led to advanced research which led to advanced development, full scale development, production, and so on into operation.

Two decades later in the mid-1960s another study, Project Hindsight, reviewed the predictions and came to two conclusions that seemed to be outright contradictions of the original presumptions. As such, they were vigorously contended because, among other implications, they seemed to undercut long-standing justifications for basic and applied research.

The first and strongly factual conclusion of the Hindsight Study was that even Foresight's experts fell far short in many areas, including television, guided missiles, nuclear weapons, and extensive military use of satellites. One could argue that this outcome should not have come as the surprise that it did. After all, any study attempting to predict 20 years ahead in the 1920s to WW II in the 1940s would have suffered the same fate. Hindsight and Foresight were simply reconfirming the insight, to

Expect the unexpected.

The second conclusion was, and still is, much more controversial: no new defense system in the intervening 20 years since WW II could be credited or even traceable to *any* of the many intervening research programs or fields. That is, 20 years of individual research projects had not resulted in any jet aircraft, satellite, ICBM, gunship, or nuclear submarine systems. Needless to say, this second conclusion did not result in great enthusiasm for funding basic research at the national level or for increasing what little basic research there was being done in the commercial sector. But even more serious than the factual conclusion itself was the lack of understanding of *why* it said what it did.

As it has now turned out, Hindsight, in connection with reaching the second conclusion, observed an essential clue to its underlying cause. Every major new system required *many* new developments in *many* fields to produce the observed results.

For example, a space system represents far more than a single research project in rocket propulsion, electronics, materials, or communications. At the same time, without developed research results in each of the necessary fields coming together synergistically, new systems and their emergent capabilities would not have been possible. Therefore, although no single field of research by itself made possible a dramatic new system, the totality of all fields of space research and development did. Properly managed, they provided the choices of results so important to the totality of space systems needs.

Today, half a century after Project Foresight and knowing much more about complex systems, we should probably modify or at least clarify the two Hindsight conclusions. The first conclusion would be restated as follows: we should *expect* the future to be unexpected. The second conclusion would now be restated, less controversially, as

New capabilities are most likely to emerge from those systems
that are able to justify the timely development
of a selected set of research results.

This conclusion implies the existence of the following:

- An array of basic research results in many fields from which to choose sets of results that could be developed further — in essence, the justification for broad-based basic research.
- A small set of well-architected prototypes in critical applications as the mechanism for synchronizing the choice and development of those sets of research results — in essence, the justification for prototyping and other advanced developments.

The first implication could be described as the "push" and the second as the "pull" for technology management. The first is not really surprising. Regardless of both studies, basic research has had good federal funding for more than half a century. The second implication appears to have been short-changed, or at least as articulated in this particular way. The missing element in both studies seems to have been recognition of the need for well-architected proto-systems to synchronize the development of selected research and engineering results.

A practical example of one such research development synchronization was that of the 1950s era JPL planetary radar; in effect, a direct prototype of the Deep Space Network (DSN) of tracking and data acquisition stations intended to support spaceflights to the planets and beyond. The radar, by sending and receiving a signal to and from the planets well before any spacecaft would arrive, made sure that no technology would be late when a similar signal was sent to the spacecraft at the later time. The timing milestones for synchronization were the few weeks every 18 months when Earth and Venus were closest so that radar reflections could be at their strongest. Synchronized research was carried out in the related fields of signal coding, transmission, maser reception, phase-lock tracking, computer-aided design of antenna structures in hostile environments, orbital mechanics, and others in order that, at each closest approach, all elements would be "go" with no "wait for me." As hoped and planned, the radar measured the distance from Earth to Venus with increasing precision at each opportunity.

Results increasingly justified making detailed performance commitments to the spacecraft designers more than 10 to 15 years before the spacecraft would arrive at its target planets. (One mission took 12 years of flight and 7 years to design and build. Critical to the design was the initial commitment by the DSN to a specific capability 19 years before the spacecraft arrived at the planet.) The promised capability on average was more than 100 times greater than what existed at the time of commitment, a promise dependent almost entirely on synchronized developments in more than half a dozen fields for almost 40 years. It should be noted that the promises, if they came up low, would result in a corresponding decrease in spacecraft performance. If they came up high, they would preclude otherwise achievable results. Thus, a promise that was either too high or too low would be equally damaging to the mission. In the end, the promise of a factor of 100 improvement within 20 years was met to within a few percent. The DSN was never late.

Quite clearly, had any of the broader research program's results not been developed in time, the endflight projects could not, or would not, have flown. The conclusion for the purposes of this book is that synchronized R&D in multiple fields is not only possible, it is essential in some high-tech applications.

Now back to organizations and change. The importance of synchronized development is particularly important at two different times. During evolutionary change it keeps product lines advancing from one block to the next.

During radical change it provides more orderly dispositions of related efforts from pre-change to post-change. Rather than having a scattering in time of unsynchronized developments to deal with, the organization will have time-synchronized sets of related efforts as viable options regardless of when the change occurs.

Summary

This chapter has focused on the characteristics of complex systems and organizations that make them unique and distinguish them and their methods from those based more on science and engineering disciplines. In particular, problems are ill-structured, replicable measurements are rare, and the results are not always as expected in complex organizations. Several such problems are discussed including overlapping regions of interest and responsibility, unexpectedly chaotic behavior, and synchronized research and development of architected prototypes.

As will be seen in the chapters to follow, many of the approaches, techniques, and tools for developing systems and organizations are different as well. *That*, at least, should be expected.

Notes

1. From the Introduction to Part I: "Complexity is defined as the degree of intricacy of a system so interconnected as to make analysis impractical. The more interrelationships in the system, the more complex it is." For organizations, this would roughly correspond to more than 7 ± 2 interconnected units, effects in any one of which would deeply affect all others.
2. The words "parts," "elements," and "units" are intended in this text to mean things big enough to have significant internal structure. Automobiles have "parts," systems have "elements," and organizations have "units," but in this book all three will mean the same thing from a systems point of view. "Components" will be used to mean things like nuts, bolts, resistors, a piece of cable, etc.
3. [R&M 97 121-130] The other perspectives are purpose or objective, behavioral or functional, performance objectives or requirements, data and managerial.
4. For more on feedback, nonlinearities, and chaos, see Rechtin and Maier, *The Art of Systems Architecting*, CRC Press, 1997, 63-66.
5. In common usage, "unintended" is the generally accepted term, but it usually implies negative consequences. "Unexpected" is used here to imply positive ones as well, e.g., an unexpected "bonus" from using a complex system in an unexpected situation.

chapter two

Emergent capabilities and values of systems and organizations

No system or organization can long survive without a viable purpose.
The test of a good architecture is that it will last as an enduring pattern.
The purpose of an organization and each of its organizational levels
is to create capabilities and values in addition
to those of their separate elements.
Value is in the eyes of the beholder.

Introduction

Chapter 1 introduced classes of system and organizational problems for which applied science and mathematics were inherently insufficient.

This chapter introduces the concept of systems capabilities and values which "emerge" only after the system is assembled and operating. By definition, emergent system capabilities do not exist alone in any of the separate elements.

Three emergent capabilities will be presented in some detail: quality, knowledge, and core competency. They are of particular interest to excellent organizations because each interacting with the others determines in large part what such organizations can or cannot do.

Consistent with the systems approach, the context of the chapter remains that of the organization as a whole. Value judgments and technical decisions in subordinate elements are made based on their effect on the properties and functions of the system as a whole. This is not to say that the elements are not important. Indeed, they "enable" the system to exist at all. But they cannot, of themselves, individually produce an emergent system behavior. For example, the human heart, lungs, brain, muscles, and so on cannot individually produce life — the behavior emerges only when all elements are functioning properly and together.

The meaning of emergent value in organizational performance

"Value" as a general term has at least a dozen definitions, but the one most useful here is that of the economist. Economists define value, or worth, as what someone (else) is willing to give up to get. It inherently is partly objective (cost) and partly subjective (desires).

Value, or worth, is in the eyes of the beholder.

Clearly, the *value* of a given system capability, as opposed to the cost of producing it, is different depending on whether the stakeholder is a buyer, a lender, a seller, or a competitor. The price of a house, for example, is not determined solely by what the seller might desire but also by negotiations, taking into account what the buyer is willing to pay.

The transportation value of a particular automobile or truck clearly includes more than its negotiated price. To a wage earner it may include a broadened choice of job opportunities. To a teenager, as parents well know, it is freedom. To a traveling salesman, it is the potential size of his territory.

The value of research, of publications, of resource management, and many other functional results similarly depends on who is judging its value and by what criteria.

Price negotiating, therefore, can be more than a zero sum game in which what one side gains the other loses. In effective negotiating, few parties gain everything desired, though everyone gains something. The nature of that "something" no doubt will be different for each party. The end game — much as it is in architecting and design — is one of fit, balance, and compromise until a solution satisfactory to all conerned is reached.

Value can be produced in many ways. It can be in the form of a product created by processing inputs and materials into product outputs. Existing processes can be made more efficient or cost effective. Productivity of an organization can be increased by reducing duplication of common support functions such as finance, personnel managment, and legal affairs. Profit can be increased by making systems smarter.

And, as reflected in the theme of this book, value can be generated as new and unique system functions are created by organizing otherwise separate elements into a system.

The family car case

A favorite illustration of an emergent capability is that of the family automobile. Envision your automobile in the middle of the driveway. Now imagine asking a first-rate mechanic to disassemble it down to its nuts and bolts, taking care to lay them out in the driveway without damaging any of them and neither adding nor subtracting anything from the total. Everything is still there — tires, wheels, seats, windows, pistons — everything. What do you *not* have? The car cannot run and, therefore, you do not have transportation. Transportation is a higher level capability created solely by the system as a whole. None of the parts can produce it separately, though certainly all are required. Physically, the only differences are that the parts are now connected, producing transportation and its added value. This connection is more than plugs and sockets, it is the *relationships* that add the value.

In the automobile of year 2000, all but a few percent of the older mechanical, hands-on relationships between driver and automobile will have been replaced by electronic linkages of microprocessors and software. The design, or architecting, of these relationships largely determines how, and how well, the assembled automobile can perform. Most automobiles now have computer-controlled shifting, steering, brakes, fuel injection, and air conditioning. Easily predicted are collision and obstacle avoidance and automatic blind-zone mirror adjustments for lane changing.

Clearly, the system capability or function (transportation) has been enhanced by such changes. Driving is both safer and faster, less tiring and more worry free. Cars and trucks are less expensive in constant dollars and better equipped for more purposes.

But wait! You still do not have transportation, at least beyond your driveway. External to your automobile must be roads, rules of the road, law enforcement, licenses, gasoline and oil, gas stations, sales and service, maps, and automobile insurance. The family car, in truth, is only one of many nearly autonomous systems required by a transportation megasystem. That level of system creates value of its own — the mobility of a widely dispersed population.

Highways can be improved — more forgiving of driver error, better marked and signed, fewer intersections, and more traffic control at access points. Even more megasystem functions are being proposed — intelligent highways for traffic control, satellite-provided car and truck navigation and location. Now if we could only automatically "ground" drunk drivers, improve the separation of heavy and light vehicles, eliminate railroad crossings, etc.

Which, of course, leads to another whole system — the management of the entire megasystem and its parts, including the automotive industry. Rather than attempt to cover the entire industry, this book later will focus on only one of its emergent functions — quality — as one more perspective on what companies can and cannot do.

Emergent corporate-level capabilities

What capabilities can corporate-level organizations, as such, develop from within their structure? And what can't they develop? Can the organization combine point-of-sale with inventory management to eliminate internal buyers? Yes, Wal-Mart did it. Can it combine dealerships and volume purchasing

to make the economics of automobile sales much like that of such commodities as major appliances?" AutoNation Superstores, a part of Republic Industries, seems well on the way to doing so. Can it build both airplanes and
ships? General Dynamics did. Can it add computers and calculators to
instruments to produce programmed testing systems? Hewlett-Packard has.
Can it be easily privatized, deregulated, or re-regulated? Experience with
remote sensing satellites, weather satellites, and public utilities would seem
to say, "With difficulty."

Underlying all these domain-specific examples is the question, will com-
binations of various elements really add new kinds of value or not? That is,
what, if any, new and unique system capabilities will emerge as a result of
organizing elements into systems, compared with letting them go their own
separate ways? In particular, what new capabilities and values are brought
out in mergers? What old functions are eliminated in deregulation? What
values in which areas are attributable to public sector management vis-à-vis
private mangement? In other words, what kinds of organizations do which
kind of tasks best? Architectural perspectives, if not answers, will be presented in Part IV. Meanwhile, on to specific emergent properties and their
values, beginning with quality.

Quality as an emergent value

The lean production story

A dramatic example of system functions emerging from a system as a whole
is the story of lean production. Production, as most readers know, was
revolutionized when in the early 1900s mass production progressively superceded handcrafting. Henry Ford and his associates pioneered the process
and steadily improved it. What had been expensive and labor intensive
became affordable and labor efficient. Education changed from one of
unschooled apprenticeship to one of well-paid, grade school graduates who
now could afford to buy the products they made. Automobiles and trucks
were priced to sell to and maintained by this customer base for an acceptable
number of miles and years.

After World War II, with help from now-famous, then-rejected Americans, including W. Edwards Deming and J. M. Juran, Japanese automakers
decided to enter the world market with better, but still affordable, cars. At
the time, Japanese bankers were unimpressed by their automakers' global
ambitions. Among other reasons, worldwide sales and service didn't exist
and without them repairs and warranties would be expensive. In a sense,
there was little choice for Japanese automakers but to make better cars for less.

Deming and Juran were convinced that better cars would make a profit,
rather than just cost more, and they so convinced their Japanese clients. A
number of techniques were tried. Taking to heart the descriptive heuristic
of Murphy's Law:

If it can fail, it will.
The Japanese added their own prescriptive one:
If it can fail, then fix it first!

Deming's most effective advice might be summarized in his own set of heuristics:

Tally the defects
Analyze them
Trace them to the source
Make corrections
Keep a record of what happens afterwards.
To which can and should be added,
And keep repeating it.

Beginning in 1959 and for almost 20 years, Japanese automakers had tried to break into the American market without marked success.[1] Toyota tried first, selling cars of rather poor quality in the American market, but their performance and value were far from competitive. Pioneered by Eiji Toyoda and Taiichi Ohno at the Toyota Motor Company [WO 90], Yoshiki Yamasaki at Mazda [HAY 88], Genichi Taguchi at Nippon Telephone and Telegraph [PH 89], and others, Japanese companies then tried a number of approaches, including worker empowerment, quality circles, automation, total quality management (TQM), and just-in-time inventory management (JIT) [PH 89].[2]

Very few manufacturers elsewhere paid much attention. After all, even though improvements were notable in special cases, they were relatively small overall — a few percentage points at best for any individual approach. Even so, this quality-aggressive approach differed considerably from what then was the American automakers' strategy of planned obsolescence.

To them, quality cost *money.* It was not yet apparent that quality could make money if the quality was high enough. On the contrary, trading off the cost of increased quality methods on the production line for diminished warranty costs was barely a "wash," and a wrenching change in the rules for little net gain in sales. No one, probably even including the individual Japanese automakers, had yet understood what might happen if all the approaches, plus some high-level policy decisions, were used to create a production system in which quality was an emerging value.

In the meantime, Japanese manufacturers also came up with a number of important, experience-based, simple, practical manufacturing-line strategies, two of which were:

The five why's
and
Everyone is a customer, everyone is a supplier.

The first strategy directs failure analysts to ask, "Why did it fail?" and then, "Why did that fail?" and so on for a total of five times until the root cause had been singly and unambiguously identified. To see its power, if one assumes that at each question there might be three possible explanations, then five why's would have explored almost 250 possibilities! It was and is a powerful diagnostic tool for raising quality from *low* to *good*, but, of itself, not from *good* to *high*. And if the cause proved to be a design flaw, its effects might not be apparent for years.

In the second strategy, each worker on a production line inspected each unit arriving at the workstation and had the option — indeed, the *responsibililty* — of rejecting it and "pulling the cord" to stop the manufacturing line (a *verboten* on a conventional mass production line) until the unit was fixed. But, at the same time, that worker's output was looked over by the next worker, and so on. Quality became everyone's business because no one wanted to accept a poor product and especially not to deliver one. It became a matter of self-interest and pride not to have the next "customer" pull the cord. This strategy, like the first one, worked best if the quality was already good. Otherwise the line might be stopped much of the time compared to the "fix by replacement from inventory" strategy of most mass production lines.

By the mid-1980s, the Japanese production lines and their products could be classed as *very good* but still limited in their ability to capture world market share. In fact, the effects on the bottom line of profits depended more on currency exchange rates than specifically on qualilty.

Then a remarkable thing happened. In a very short period of time, as little as 5 years, everything seemed to come together and the results that emerged were dramatic — across-the-board, continuing increases of 20% or more in profits, production yield, design life, sales, market share, inventory reduction, time to market, cost reduction, and above all, high quality and access to global markets — *everything*. These radical improvements in overall quality and value were the direct result of a remarkable development in automotive production called lean production [WO 90]. One way to understand why this happened is to look at the Japanese production process, lean production, as a system.

Figure 2.1 depicts a mass production process, slightly modified to fit a systems architect's perspective. In Figure 2.2 a lean production process is drawn to emphasize the major changes in policy, the establishment of strong action–response relationships with suppliers, the aggressive post-sales contacts, and the three system "feedback loops," all of which together made lean manufacturing a practical reality. Notice how the suppliers, now responding to quality imperatives by being tied into a closed action–response "loop," could help manufacturing and assembly send a higher quality product to test and certification. Notice how test and certification, given a high-quality product, could reduce its costs, increase its pass-through rate, and reduce time-to-market. Notice how the tasks of sales and service could be better focused on the customer, and so on. No wonder the whole process is

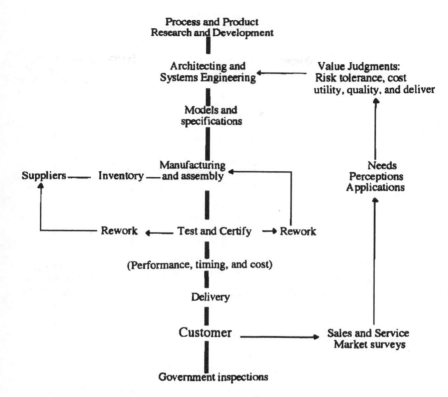

Figure 2.1   An architect's sketch of mass production.

made simpler, faster, and less expensive because the drag of "work in progress," inventory, and warranty repairs could be reduced. And no wonder so many benefits emerged all at once. To understand how many of the benefits emerged from the system as a whole, imagine that any one of the loops suddenly lost its quality edge. What would be the effect on all the rest? The effect would be a 20% drop across the board from time-to-market to profits, no matter how high the quality of the other elements might have been.

Treating lean production as a system also answers a question left unanswered by the MIT study [WO 90]. How can other companies convert from mass production to lean production? The key to the successful Japanese shift from conventional mass production to lean prduction was the policy commitment to high quality at every stage, from every supplier, and during every process. Without that commitment *first*, nothing much would have happened. One can not pick and choose elements. One can not blame the workers, the unions, the suppliers, the salespeople, even the equipments. All of these have been exonerated, one by one, except for management (company and union) which did not change the way their companies did business by making the value judgments and policy decisions essential in architecting a new production paradigm — lean production. Excellent companies in many

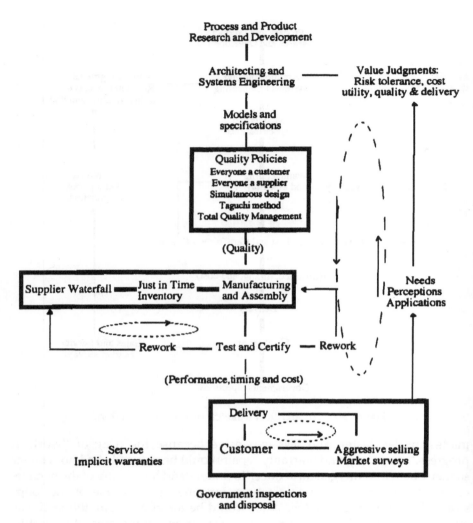

Figure 2.2   An architect's sketch of lean production.

fields, pay notice. Radical change, which is what lean production is, is not incremental adjustment of existing policies and practices. Without making these changes early, if not at first, they will fail, period.

Why one-at-a-time measures do not create emergent results

TQM, JIT inventory control, robust design, and worker empowerment techniques applied to lower quality processes would not — indeed, could not — have paid such handsome dividends. JIT, by itself, could not reduce inventory costs because the primary reason for inventory (work in progress) is to have excess parts in bins next to the production line to replace others that failed during assembly. Unless the quality goes up, the inventory cannot go down. Hence, JIT could only *shift* costs from automaker to supplier and transporter. TQM alone would become only a limited-time management initiative, as bits and pieces were chosen by managers who happened to favor them for reasons that had little to do with product quality undefined. Robust (Taguchi) design would be folded into quality assurance testing departments where its results, for organizational reasons, might have little effect on design. *Only if quality is already very high* can these methods work to make it even higher. One might say that organizational and system quality is "bi-stable." On one side of a dividing line, low quality keeps low quality in place. On the other side, high quality creates still higher quality.

The initial American Big Three response

The American Big Three[3] manufacturers, impressed with the Japanese results in market share, cost and profit results, decided to follow the Japanese lead, but with only one technique at a time. One of the first to be tried was JIT; in actuality a result rather than a creator of quality and, frankly, it was a disaster. The suppliers revolted at the shift of replacement inventory costs to them and kept on producing at the same quality level as before. A tarnished JIT was discarded. Tighter relationships with the suppliers were tried and this time with more encouraging results for the automaker, but accompanied by an abrupt cancellation of most (less qualified) supplier contracts in favor of a few (better) ones. TQM was applied to nonmanufacturing organizations from think tanks to universities where quality is very difficult to define at best and has little to do with manufacturing at worst, and then slowly folded into good management, losing its cohesiveness as a partial system.

New American suppliers to Japanese transplants in 1998

And so it went, American manfacturing results were mixed at best. There was one remarkable exception, the so-called "Japanaese transplants." These were Japanese manufacturing plants "transplanted" (conceptually, not actually) to the U.S. but staffed with American workers "out in the green fields" of rural America.[4] As told by Lindsay Chappell in an *Automotive News* report [CH 98], American suppliers to American manufacturers of Japanese automobiles (Toyotas, Nissans, Hondas), trained by the Japanese, not only converted rapidly to lean manufacturing, the parts they produced and the cars comprised of them, were at least as good as those produced in Japan — and more competitive, *even in Japan*. One of Chappell's sources, David Cole, Director of the University of Michigan Office for Study of Automotive Transportation stated, "You could say that the robustness of our economy is due to the fact that we brought in so much intellectual capacity (from Japan) and we are beginning to use it today." An American supplier to Toyota, Fred Soafer, made the telling comment,

It's a beautiful thing when it's all working together!

which is hard to beat as a reaction to the emergent system functions of lean manufacturing.

Astonishingly, this greatest single revolution in emerging value in 50 years was given free to the U.S. by Japan. It gave the U.S. better transportation, high-tech employment, a reduction in the trade deficit, and new and prosperous cities, all against serious opposition by the U.S. Government and manufacturers. Probably even more important, it demolished the theories that Japanese cars were better because Japanese workers were better. Not so. Both American and Japanese workers are very good, properly led.

Knowledge as an emergent value

To this point, value has been treated as it has been for 100 years — that of products, processes, people, facilities, profit, and so on. It is time to treat it differently. It is time to add something new to the conventional list — knowledge.

A historic trend

Peter Drucker, in a 1987 Lord Foundation lecture at the University of Southern California, spoke of the changing meaning of wealth over the centuries. [DR 87 49-54] Wealth used to be measured by the ownership of land and its natural and human resources. However, nations still go to war over seemingly minor changes in borders and ethnic minorities still fight for independence. Professor Drucker presented his short history of economics into the near future with the following insight:

> *Today, wealth is no longer best defined as ownership of land, goods,*
> *capital, or labor. It is knowledge and knowing how to use it.*

Drucker supported his observation by noting how quickly new companies and countries become wealthy by creating, acquiring, or being given "know-how" knowledge and modern technology and then exploiting them. Not surprisingly, the developing countries caught on quickly to the opportunities that were presented, leapfrogging knowledge and technology that had become obsolete. But what defined the knowledge they wanted?

Knowledge and its use, like other values, is difficult to measure. Is it books, libraries, lines of code, designs imbedded in products, Web pages, skill, or years of schooling? Or is it understanding, expertise, and wisdom? Or all of the above? One indication is the relative market value of the software giant Microsoft ($212 billion) compared with that of the then Big Three American automobile manufacturers ($150 billion total).[5] Another indication is what foreign customers wanted most.

In the late 1980s, the Chinese Government, with the approval of the U.S. Government, began negotiating with American aerospace companies for purchasing communication satellites, but in a very special way. One satellite, as expected, was to be put in orbit. But a second one had to be provided to them on the ground, in China, along with all its documentation and as much know-how as the American companies were willing to provide. For whatever reason, very few additional American satellites were to be ordered. Suffice it to say, the Chinese had learned how to build and launch their own.

As another example, Secretary of State Henry Kissinger at one point offered American technology to other developing countries, only to realize subsequently that it was not his to offer. It was owned and closely held by the American firms that created it. It was an asset too valuable to be given away for the price of its documentation. As this case turned out, the American companies licensed (sold) the technology with U.S. Government approval.

And, in a final example, the intelligence community late in the Cold War years did a study of technology transfer from one superpower to the other and concluded that (1) transferring physical things resulted in only a temporary benefit which soon disappeared, (2) transferring "know-how" documents such as nuclear weapons designs resulted in a longer lasting step upward, but (3) the most effective way, by far, was the transfer of a small number of people who had the necessary knowledge and knew how to use it.

In a related case, when the country of Iran shifted from being a U.S. ally to being an adversary, there was real concern in the U.S. Department of Defense that the first-line aircraft and weapons that the U.S. had supplied the earlier regime might be turned against the U.S. in the future. For various reasons, this possibility never occurred. In point of fact, within about 1 year after the Iranian Government changed, the American weapons, no longer being supported with parts, spares, and training, could not be kept operational. As the author understands it, the result was much the same with the Soviet Union and its departed allies.[6] Knowledge and technology evidently require continuous refreshment.

Hewlett-Packard had a telling experience which confirmed the first of the intelligence community's conclusions. In the 1950s it transferred a complete manufacturing plant for making vacuum tube voltmeters to a Warsaw Pact country. Americans visiting that plant some 30 years later found it much the same, making the same voltmeters, now long obsolete elsewhere.

But by far the greatest single transfer of knowledge was from the automakers of Japan to their suppliers and manufacturers of Japanese autos *in the U.S.* So successful was this transfer by Toyota, Honda, Nissan, Mitsubishi, and others that their "transplants" now make up 22% of the U.S. market. Surprisingly, perhaps, the transfer was not of machine tools and automation, but of a few managers, management policies, and techniques — knowledge accumulated over 20 years of trying and the lessons learned in "lean" manufacturing.

Creating, applying, and controlling knowledge has now become one of the central purposes of organizations at every level. It is one of their greatest assets, a product they can sell or transfer as they think best. As intellectural property, it is accorded the same legal protection as most other property, and is fought over almost as vigorously. And, yet, it is not a "thing."

On the one hand, knowledge is very difficult to contain or own indefinitely, even the most secret and most heavily secured. On the other hand, unless refreshed and updated, it loses its value relatively quickly.

On modifying Drucker

These cases, collectively, suggest that Drucker's conclusions perhaps be modified to:

> *Today, wealth is no longer best defined as ownership of land, goods, capital, or labor. It is* new[*] *knowledge and knowing how to use it.*
> (Steven Carlton 1998)

New problems

The rapid increase in knowledge has resulted in new kinds of problems; in particular, those of organizational confidentiality and personal privacy. The extraordinary browsing power of the Internet, the World Wide Web, and local area networks (LANs) for accessing, if not stealing, company and private files has raised serious concerns in organizations accustomed to treating organization charts as the only "official" channels. Those channels were enforceably controlled as to what information could and should be sent on them. Now there are channels from everywhere to everywhere else. True, everyone should now be able to "get the word," but from a managemement perspective, not everyone should get much less redistribute every word on every subject. Some companies have insisted that access to the new networks be sharply curtailed, some have shut it down altogether. But because knowledge access is almost impossible to police and is unquestionably valuable for joint problem solving, trying to deny access has seldom succeeded. Legal scholars, as should be expected, have begun debating the constitutionality of denying unfettered access to the Internet, an aspect of First Amendment freedom of speech not even contemplated a decade ago.

Security and privacy likewise are confronted with new problems. Breaking and entering to obtain information can now be done, almost undetectably, and from the privacy of one's home. No longer are the methods those of President Nixon's "plumbers" who left the door open. Now we have hackers who can enter by wire.

New forms

Knowledge in many ways is quite unlike other forms of value. It can be used to create more of itself. Software development is, in effect, an expanding

---

[*] Author: At least to the user.

spiral of learning; whereas hardware development is a linear sequence of steps, a "waterfall," to a known end product.

Transferring (communicating) information up and down organizational levels can severely distort and attenuate it, unheard of in hardware. The attenuation process is a perfectly understandable one. A 2-hour top-level meeting is held to discuss a new policy. All participants are told to inform their subordinates of what happened. Each does, taking perhaps 12 minutes each — and certainly not the original 2 hours. The listeners are asked to inform their subordinates, which they do quickly, the policy appearing to be just one more order from above. And so on. Each "summary" cuts the volume by a factor of 10. Each "pass the word" transfer distorts the original meaning depending on who is talking and who is listening (or not).

Or, going in the opposite direction, a meeting of first-level managers reaches conclusions that need to be interpreted and summarized for the next-level supervisor who interprets the gist of them to those above.

This distortion and attenuation effect is not unique to companies. It also afflicts military chains of command. Needless to say, it breeds a certain mistrust, if not outright cynicism up and down a chain in which a four-star general is eight levels away from a second lieutenant and still more from a private on the combat line.

Various observers have tried to quantify these curious properties of knowledge and its transfer. Two such quantifications are

> *Communications suffer a loss of a factor of 10 for every*
> *management level they transit.*
> and
> *The time it takes to reach a final decision doubles with each added*
> *approval required.* [RE 91 283]

So don't be surprised that in a seven-level organization only a millionth of the information generated at one end makes it to the other end. Perhaps that calculation is just tongue in cheek. But there is a serious side, too. "Lack of communication" is first on everyone's list of management deficiencies. Arguably the wide-open communications of the Internet may help, but, to do so, a completely different management environment will be needed, one that trades off access, credibility, privacy, security, timeliness, creativity, ownership, and control of knowledge.

And don't be surprised by the second insight and its corollary that some decisions take so long that the action proposed originally dies. It is not unusual for a dozen approvals, or 3 years, to be required in socio-political-technological issues, even though the first one took only half a day. The original observation was for Army tanks where the approval process for a new tank in the 1950s took only a year or so. By the 1970s it took 10 years. With additional governmental oversight, it now takes 20. Needless to say with all the changes, it is no longer the the same tank described in the proposal begun 20 years previously. Hence, the common wisdom that if a proposal takes more than 3 years (or 12 approvals), forget it.

Lessons such as these are only rarely caused by stubbornness, stupidity, or animosity — the standard list of suspects to bash. There are, and always will be, structural constraints on the ability of any established organization, society, or country to respond to major change. These constraints are the result of a need in human lives for stability, safety, certainty, and security, and they are unlikely to change. Unfortunately, they are one of the main reasons that excellent organizations cannot accomplish all that they might have wished and believed they could.

Proprietary and licensed knowledge

One of the greatest competitive battles in the information technology field has been between one company, Apple Computer Inc., that attempted to keep its computer operating system proprietary, and another, The Microsoft Corporation, that made its operating system available to everyone under license. Apple, with a remarkably user-friendly product at the time, reasoned that keeping it proprietary should stop others from copying or "cloning" its features and reducing Apple's then sizable share of the computer hardware market. Microsoft, on the other hand, reasoned that by letting everyone, under license, use its proprietary product, *it would expand the total market* and, consequently, share in what turned out to be an explosion of users and applications. The first approach resulted in a steadily decreasing market share and absolute size. The second approach reputedly made its chief, Bill Gates, the richest man in the country. [MO 93]

The lesson here is not necessarily which company made the correct licensing decision at the time — a small detail in the overall scheme of things — but what it says about knowledge as wealth and value. Metaphorically,

> It is about as hard to stop the flow of knowledge
> as to stop the flow of water.
> The best one can do is to channel it, and
> even that can be overwhelmed.

Whether the "pull" is gravity, need, or profit, it will flow. The sources keep sending more of it all the time, forcing it to find other channels, overcome extraordinary obstacles, and rest only temporarily. Apple Computer tried to restrict the flow of knowledge, Microsoft went with it and managed to share in the value it created as it flowed. Knowledge, after all, is not a static *object*; in many ways, it is more like a fluid. It is probably not just a coincidence that we describe both in a similar manner — as a body of knowledge or a body of water.

Excellent organizations where their wealth is based on people and their knowledge are confronted with a double problem. Both can flow out the door at any time. Neither can be contained indefinitely, nor should be. Rather, use them well while you can and take whatever advantage you can as they flow in, by, and out.

Educating to advantage

Drucker's insight, that wealth is (new) knowledge and so on, is more than that. Knowledge alone is of little value, Drucker implies, if an organization doesn't know how to use it. It is as useless as a locked library. Knowledgable professionals in an organization are of little value if the organization doesn't or can't use them. Conversely, individuals with critical "know-how" can be priceless. As Bob Spinrad of Xerox put it in thinking of the future options for Xerox, "Systems architects will have more influence on the future of a company even than CEO's.[7] Pay whatever it takes, but get them and keep them!"

Understanding the value of knowledge says much about what an organization can or can't do. It pinpoints the necessity of a company that is investing in purposeful training and education of its staff if it is to have options to change in the future. For example, one of the better clues for deciding whether a company is positioned to change or not is the nature of its training and educational program for its employees. If its policies limit educational possibilities to those of immediate worth, its ability to change will certainly be limited. If it encourages education at the edge of its domain, it creates the option of moving knowledgably in that direction. External consultants are no substitute for in-house knowledge.

In short, knowledge and how to use it has added a new meaning to value and to the systems that create it. New knowledge must be added to the list of functions that emerge as elements are brought together into systems.

Core competencies as an emerging value

What is the value of a company or agency? It is clearly more than the price that a company's facilities would bring on the market. In stock market terms, it is tied to what a buyer thinks it might bring in the future, which is part speculation and part calculation. But these assessments leave out a great imponderable, the core competencies of its people.

It is often said by chief executives that "our people are our greatest asset," acknowledging that a company also includes other assets: its facilities, its accounts receivable, its land holdings, its goodwill, and so on — factors of major import in farming and manufacturing. In knowledge-intensive organizations, on the other hand, a more realistic statement might be, "Our company *is* our people." If all the company's physical plants were destroyed or all of its people simply got up as a group and left for other quarters, the company would still exist. The long-term limits and opportunities of the company would be much the same. Indeed, in some cases they might be better. A case can be made that some organizations might be better off if they simply abandoned their obsolete physical plants for new ones. As one heuristic puts it:

In open competition,the incumbent has the encumbrances. [WI 91]

It is an insight that applies to the Big Three U.S. automakers who are faced with Japanese transplants in nonindustrial regions of the U.S., that of using new and enthusiastic workers, advanced methods, and designs developed by Japanese and European automakers.

Therefore, unless great care is taken by excellent, long-standing organizations, the problem of encumbrances will apply to them and their people as well. The amount and form of compensation (wages, salary, health, and retirement benefits) that can be offered will shape what its people are willing to do. That is to do a good day's work every day, even weekends, to deliver more than expected, to volunteer for tasks that "somebody has to do," to strive for excellence for their company as well as themselves, or to commit to a 20- to 50-year career with the company. And needless to say, what its people can or can't do, the company can or can't do.

But, more than any of these valued assets is the mutual dedication of employee and employer to the company in good times and bad. When it comes time to change, this mutual dedication and resilience to stress is priceless, impossible to quantify.

Excellent companies recognize that compensation for such voluntary dedication must be more than wages, salary, and benefits. They recognize that for most professionals, once financial compensation is noticeably above survival level and has reached a useful level of discretionary income, it rapidly loses its value. For them the most powerful driving forces in their lives are no longer financial. They knowingly forego better paying jobs in order to do what gives them personal satisfaction — professional pride in a job well done and peer recognition. They are more likely to give up family needs to move at the company's request. They are more likely to reject invitations to leave the company for another one for less than about 20% more than they are currently earning. They are far more likely to recommend the company to others, considerably reducing recruiting costs and keeping standards high. To an observant stakeholder, such companies have a much better chance — and noticeably more real possibilities — to respond to change.

But professionals, rightfully, expect the organization to give value for value received — challenging assignments, an implied stability in their employment based on their contributions, fair treatment in promotions, and a trust and confidence in themselves and their work. In short, loyalty given for loyalty received.

It is fair to say that the implied contract between professionals and their organizations has been strained during the last decade, even in excellent organizations. The causes of the strain are generally understood. In the public sector, they are global competition and the resultant restructuring in response to it. In the public sector, it is a decreasing respect for government service.

Professionals engaged in government service see this loss in the harshest terms. Joining a governmental institution has become hazardous to one's health, increasingly rejected by the very people with the skills and experience that are so badly needed. They are not applauded by the Congress and the media, they are more often demeaned.

Refusal to serve is now far more common — and understood by recruiters — than was the case even a decade ago. As late as the mid-1960s, acceptance of a proffered government position was regarded as an honor and as part of one's duty to one's profession and country. The transfer of skilled individuals from industry to government and back again was internationally recognized as a unique strength of the U.S. Government. Granting permission to an employee to take a government position was a company obligation and often included a pledge to reestablish the individual in the company on return. Such two-way transfers, for all practical purposes, are now viewed by the Congress as impermissible as conflicts of interest.

The total compensation, financial and personal, is no longer adequate to attract qualified people even at the highest levels of government. Even before the impeachment of President Clinton, exceptionally qualified professionals refused to consider becoming presidents, governors, senators, judges, and heads of agencies. It has become not worth the price of public vilification. Veterans of government service at all levels have discouraged others from taking their place. One result is the substitution of power and influence for dedication to public service in the minds of applicants, seriously constraining what the needful organizations are able to do without the appearance of personal gain.

One of the most serious casualties of this state of affairs is the opportunity for government organizations to be seen as excellent in the eyes of its own professionals. At risk are the dedication, skill, and resilience that are absolutely essential in striving to reach an otherwise impossible dream. Consider one such dream proposed by President Kennedy in the early 1960s.

Perhaps at no other time in the past 50 years have the effects of dedication and resilience on realizing a dream been as apparent as in the Apollo flights to the moon. When that project was begun by NASA as a result of President Kennedy's challenge, a calculation was made by its architecting team,[8] assuming all elements from propulsion to rendezvous and life support were done as well or better than ever before, that 30 astronauts would be lost before 3 were returned safely to the Earth. Even to do that well, launch vehicle failure rates would have to be half those ever achieved and with untried propulsion systems. Electronics and communications would have to work in conditions of vacuum and radiation never before encountered, and so on. The only possible explanation for the astonishing success — no losses in space and on time — was that every participant at every level in every area far exceeded the norm of human capabilities.[9] Unquestionably, part of the success was attributable to a zero-defects approach similar to that in World War II when patriotism was exceptionally strong. Another part, no doubt, came from the pledge made to a recently assassinated, widely loved president. And much came from a national desire to demonstrate primacy in a race.

Whatever the causes, such a success will not be easily duplicated. It gives pause to those who would claim, for less driven goals, *"If we can land a man on the moon, then why can't we..."* And it helps explain why only those organizations which are believed by their own people to be world class and dedicated to keeping it that way should even attempt to "dream the impossible dream." Because for others, it *is* impossible.

The author reached this last conclusion to his dismay after visiting manufacturing plants of several companies generally portrayed as excellent. With a simple insight, later validated by production statistics, one could tell which plants could and which plants could not produce future world-class products *just by looking at the floor.* Ridiculous conclusions with so little data? Not at all. Serious? You bet. One manufacturing plant made explosives and rocket propellants. Another made automobiles. One could, and did, do well. The other did not, and could not do likewise.

The clue? A clean floor during working hours proclaims dedication, professionalism, and pride. No cart tracks, no footprints, and no cigarette butts and discarded directives on a spotless white floor. No number of janitors or publicity releases can change the reality.[10]

There are other clues, of course, which can further refine the assessment. Are there performance charts at the exits from the work areas that plot not the average defects per month (the wall of fame), but the number of days since the last defect (the wall of shame)? In the former, a defect is just a small blip on a record of "excellence." In the latter, a single defect produces a dramatic crash on the chart and a challenge to pride.

Is the visitor guided through a sector by its foreman or by a security-conscious member of the public relations department? If the former, there is mutual trust and confidence between management and production chiefs.

Are there obvious provisions for the safety of the employees not only at the workstation but in the stairwells and aisles, on stairways and ladders, in the parking lots at night, on cushioned protrusions, on roving safety inspections, on well-placed lighting? If so, the company may be doing other things right as well.

Are there weekly all-hands meetings called in a central area, replete with charts showing performance compared with management goals? If so, you are in the presence of nineteenth century Taylorism, of the "worker brawn, management brain" theories of management.

In contrast, it is a special pleasure to encounter a flash of pride of place. The author still remembers a visit to a Hewlett-Packard plant, unannounced, and asking the receptionist at the front desk, "Who runs this place (that is, who is the plant manager)?" And receiving the smiling reply, "We all do!" It was subsequently no surprise to see that that plant had just swept the marketplace with a product that was more reliable, performed better, and cost less than that of its competitors and also less than the product it replaced, all during a time of sharp inflation.

The conclusion is clear enough. An excellent company can change, can extend, and can continue to excel only as much as its people, collectively, can. It can only take risks to the extent that its people will voluntarily risk their careers, livelihood, and even family dreams to achieve company goals. Taken for granted, loyalty can disappear overnight and, with it, opportunities for change.

Summary

The question of what organizations as a whole can and cannot do *uniquely* — that their units can not do by themselves — depends upon how much value is added or "emerges" solely because of actions and decisions on behalf of the whole, that is, by top management. The emergence of quality is illustrated by the lean production story in which a number of strategies were tried, with limited success, until all were brought together by high-level commitments, actions, and decisions consistent with a very high level of quality everywhere. Only then did the different strategies so interrelate that unprecedented business improvements could and did happen. They emerged only when everything fit together.

Knowledge, and especially new knowledge, is shown to be a new *kind* of value, one with very different properties. Unlike other forms of wealth, it is very difficult to measure or contain. It can be transferred electronically almost at the speed of light and yet, in the form of "know-how," it moves only as fast as the people who have it choose to move. It may be a company's greatest asset, but if it isn't used properly, it can be useless.

Core competencies are far more than facilities. Their greatest part is made up of the company's people, but only if they work together in dedication and belief in the company. Recent trends and actions by companies facing radical change have considerably diminished that dedication and loyalty at the risk of the core competency being far less than what might have been expected. Consequently, organizations may be over-estimating the value of their core competency in times of radical change.

Notes

1. When they finally did break in with high-quality affordable cars, American automakers, led by Lee Iococca with his "If they want to sell 'em here, let 'em make 'em here," demanded and received from the U.S. Government import quotas to keep them out, a tactic that almost immediately backfired on two counts.

   The first count: The limited quantities, because the Japanese cars were of higher quality, promptly commanded a 10 to 20% premium above the sales price — high enough to keep the total Japanese sales revenues about the same.

Had the Japanese automakers as a group raised their base prices by the same percentage, it would have been taken as illegal collusion. The Americans almost as promptly raised their own prices, a cynical strategy that alienated still more American consumers.

The second count: Given a choice of which cars to send to the U.S. under the quotas imposed, the Japanese targeted the highly profitable luxury car market to compete with BMW, Mercedes, Porsche, and Cadillac, and spun off new divisions of Lexus, Acura, and Infiniti, and increased their markets share and profits still further.

The third count: When the U.S. defined, in a surge of jingoism, that "American" automobiles be identified by their "national content," the once-reticent Japanese proceeded to make their cars in the U.S. with American workers and suppliers, while American automakers increasingly outsourced their car parts to foreign suppliers.

Further, the Japanese on learning of the environmental concerns in the U.S., solved the problem in their automobiles by design and technology often before new standards were established. The Big Three, unable to do the same, delayed the initiation of the new standards as long as possible. If ever there were arguments of whether quality is of value in a competitive world, or of which countries erect "unfair" trade barriers, or who pays the costs of the barriers, this appalling story should end them.

2. Mark W. Maier has pointed out that JIT had been a technique tried in the early years of the Ford Company. Of itself, it did not improve quality, but once mass production quality was achieved, better inventory methods were able to take profitable advantage of it.

3. Defined for many years as General Motors, Ford Motor Company, and Chrysler Corporation, the purchase of Chrysler by Daimler-Benz and the arrival of the Japanese-American transplants arguably calls for some new definitions. However, at the time of this story, Big Three was the appropriate designation for the three automakers.

4. Hence, the so-called "green fields strategy" as it became known.

5. Courtesy of Max Weiss, June 1998.

6. The Soviet Union had an even more conservative strategy than withholding parts. It apparently never gave or sold a Soviet weapon system that the Soviets themselves could not defeat.

7. A statement made to a systems architecting class at the University of Southern California. The students applauded wildly.

8. Chaired by Dr. Joseph Shea and of which the author was a proud member.

9. Skilled humans, as has been well documented, have about a 1% error rate when fully alert. The Apollo project had a cast of hundreds of thousands working on at least as many critical tasks.

10. One of the first things Japanese advisors suggest to prospective American suppliers is to clean up their production and assembly areas. [CH 98]

Part II

An architect's perspective
of the world outside

Establishing external interfaces

No man is an island. (John Donne 1572-1631)
No organization can be completely self-contained.

Introduction

No matter how large or how complex a system or organization becomes, to survive it must deal with a still larger and more complex environment around it. Part II is concerned with two of these external worlds, the marketplace and the government; that is, with the private and public sectors. Although there are exceptions, it is in the interest of all concerned that the interfaces between them help each perform its role understandably.

The buyer needs a healthy seller and vice versa. Competitors need a fair, if not level, playing field, free of distractions and uncertainties. The government and the governed need a delicate balance of power, neither able to crush the other. For many reasons, all parties want their complex interactions to be understandable and workable. One of the most successful approaches to this need to work together are the insights:

Simplify. Simplify. Simplify.
Or the more specific,
Keep the interfaces simple and unambiguous.

Different organizations, different languages

There is a very good and natural reason for simplification. Each organization has a different culture, a structure of beliefs, and behavior patterns and practices. It is no wonder that even apparently similar organizations have difficulty in understanding each other. In fact, different organizations speak different "languages" — and always will.

One of the author's favorite questions as a professor is to ask a class for a show of hands on how many languages each student knows. One? Almost everyone. Two? In California, about half the class. Three? A proud handful. Four? One, usually European, greeted with great respect. Then I tell them that each knows at least 15, though each student will know a different set. Software literate students catch on first. Fortran, Lisp, C+, C++, Basic.... Then I tell them not to forget one of the most powerful and universal, algebra! What about governmentese? Companyese. Educationese.... But how many could understand all of them? None. As Professor Theadore von Kármán[1] used to joke, but with his usual insight:

There is only one lingua franca, one universal language, and that is Broken English.

Unfortunately, Broken English, much less conversational English, is much too imprecise to be a language for defining interfaces.

Yet, there is a practical reason for this babel. We conceive, modify, and use languages to solve problems and express ideas, many of them quite abstract, some of them even deliberately ambiguous. Different problems and ideas *require* different languages. Some ideas and expressions literally cannot be translated from the original languge to another; they are too dependent upon the overall history and experiences of the culture that created them. Try translating "It ain't worth a Continental" into Japanese (and back).[2] The double translation would probably be about a well-known American luxury automobile. And so it is with technical interfaces between propulsion systems, Internet, electronics, fluid dynamics, even among software languages. C^2 means command and control, communications and computers, or 100^2, depending on the context. Now try domain-specific terms like redundant, rebooting, rationalizing, or project (making sure to accent the right syllable and pronounce the "o" properly).

There are times when the *intensity* behind the words can mean as much as the words themselves. For example, a normally quiet client, frustrated by a suggested possibility that he or she is not right but doesn't know why, may suddenly say, "I hate it! Really, I do..." The insightful architect knows that:

"I hate it!" is direction. [RU 1993]

because, if no one pays attention the next direction is likely to be "Stop it!" and that can terminate not only the suggestion but the project itself. "Hate," you see, is very seldom said quietly.

Is this "language" problem all that significant in what organizations can do? Yes. In the author's experience, it was at the heart of a serious breakdown in relationships between a major contractor and the government. Two excellent organizations were arguing past each other. A third party was brought in who spoke both languages, governmentese and companyese, and in about

3 months the breakdown was repaired and a new form of normality continued thereafter. Without any question, the language problem had determined
not only what could be accomplished by the contractor, but also by the
government and, of still greater importance, by both working together.

The different meanings of the same words, but in the different languages,
are not likely to change, lacking any real incentive to do so — and very
strong disincentives not to. Thus, the simpler the interface the better, because
its function can be better understood by the affected organizations and the
number of complex interchange issues kept modest. In the software arena,
open architectures can help decouple the separate research and development
efforts of the contributors. Interfaces can be moved such that more or less
design, engineering, administration, or production can be undertaken inside
or outside a company's boundaries. Reports can be simplified, responsibilities clarified, decisions made or delegated, testing relocated, special production lines established, command chains modified, and so forth. Generally
speaking, the less detail required by one organization from another, especially on matters involving the latter's own internal supervision, the easier
it is to deliver its products on time, on cost, and to acceptable standards of
quality. The more the fundamentals are understood across the interfaces —
things like quality, timeliness, and fairness — the easier.

Two other more sophisticated heuristics for establishing interfaces also
are understood, particularly in software and management contexts:

> *Choose boundaries with minimal communications*
> *required across them.*

> Or equivalently,

> *Choose boundaries with low external complexity (coupling)*
> *and high internal complexity (cohesion).*

These insights have been used by CalTech to good effect in the detailed
design of neural network microchips. By using the principle of minimal
communications, chips were designed that required less surface area, had a
more "elegant" (simpler) appearance, and consequently operated at a significantly greater speed. In effect, instead of separating the functions of
memory and computing, they were combined into relatively complete,
microfunctional units which then communicated with each other. The organizational equivalent is apparent in the heavy delegation of authority and
responsibility into the product divisions of Hewlett-Packard. Everything the
headquarters needed to know of a division's financial or product status was
transmitted, every evening, on a single teletype line. There was little need
to send financial or product information from division to division.

Another way of expressing the same insight is to keep complex issues
inside the company's boundaries and *not* sending them to relatively uninterested or less-informed outsiders for comment or direction. An egregious
example of a violation of this guideline was the request by one particular

government agency to its perspective contractors that *they* suggest to the government what the agency should do! The result, understandably, was a scramble by the contractors to find out what the agency wanted to hear. Architects, naturally, blanched at the idea of making the client's value judgments, one of the strongest prohibitions in architectural practice.

Efficient establishment of interfaces

Because organizations usually have a number of levels and because some units may affect a number of others "outside," dozens of external interfaces may come into being as the organization is formed. Attempting to define, fit, balance, and compromise all these interfaces in detail would place hopeless requirements on any architect. Fortunately, all interfaces are not alike. Some are well enough understood that they can be dealt with later. Some may be simple extensions of others. Still others may have a history of trouble. It is the last of these, the troublemakers, who lead to one of the most important insights in organizational design, the so-called "misfit" heuristic:

> *The efficient architect, using contextual sense, continually looks*
> *for the likely misfits and redesigns the architecture*
> *so as to eliminate or minimize them.* [AL 64]

For example, engineering departments, speaking closely related languages, get along fairly well with other engineering departments, but less well with administrative ones. Yet, all must work very closely together on such issues as tort liability, affirmative action, union negotiations, and safety regulations. An efficient organizational architect's time, according to the misfit heuristic, would then be most efficiently spent by bringing selected engineering and administration departments together on their own specific disagreements, rather than trying to sort out internal engineering or administrative turf questions. Those last two sets of relationships would probably be worked out on their own.

As a technical example, the architect of a network will be much more efficient if focused on the interfaces between software architectures — known from USC research to be inherently incompatible — than on the interfaces between hardware sets, which long have been standardized.

> In other words,
> *Do the hard parts first.*

Interfacing with the outside world

The reason that the interfaces with the outside world are so important is that they are where instructions are introduced, where products are delivered, where value is exchanged, and where support is provided. It is the boundary between what is judged to be important to delivering system performance and what is not. Making that judgment call of exactly where that boundary should be drawn is one of earliest and most dangerous decisions in all of architecting. As Bob Spinrad would say,

> *All the* really *serious mistakes are made in the first day.*

Deciding, in the first day (so to speak), what will be and what will not be considered as the organization's responsibilities, is crucial in determining what it subsequently will be able to do or not do. Too often the rationale behind outsourcing what later turns out to be an essential element is that it won't affect the design. Another common and dangerous rationale is that the outsourced element can be considered to be stable and unchanging as not to need close attention and continued supervision. And when the roof falls in, "it's somebody else's business, not mine."

Whatever the reason for exclusion of an element, excluding it will mean losing much if not all control of it. Loss of control, if nothing else, adds sometimes unknowable future constraints on internal design. The tail can wag the dog, so to speak.

The interface with the competition

If misreading a client's meaning or excluding a key element from the architecture is serious, then ignoring the reaction of a competitor to a new competitive product is apt to be a disaster. A competitor can retaliate, killing the new product, by even suggesting that a still better one is in the works. And then there is the more general reaction. If a new product becomes wildly successful, it will attract other companies, encouraging their entry into the market. If armor is added to a tank, for example, the opponent's reaction may be a new armor-piercing shell, developed in half the time of the tank. If the Americans introduce quotas on Japanese automobiles...but you have heard that story.

The lesson learned by too many excellent organizations that believed they had a world-beating product, civilian or military, is

> *For every competitive system there will be*
> *at least one countersystem.* [RE 91]

This heuristic is remarkably broad in its application and scope. It applies well beyond such technical cases as defense systems, gambling games, and sports in which head-to-head, mano-a-mano, combat is expected. Those engaged in sports know better than to underestimate their opponents. Nonetheless, many in organizational combat, when proposing a particular measure, forget until too late that a countermeasure is highly likely to appear sooner or later and, therefore, that its reality will never come up to its promise.

Two gladiators in the same arena

The conventional model for describing a competition between organizations is to show each side of a combat line. This might be called an "us" vs. "them" architecture. Examples include the U.S. vs. the Soviet Union, Microsoft vs. Apple, The U.S. Air Force vs. the U.S. Navy, all competing for funding in time of peace. In this model, the objective is victory of one and the defeat of the other.

But there is another model, one which places all the competitors in the same arena or "megasystem," with its still larger purpose of controlling the intensity and outcome of the competition. In some cases this can result in a "two scorpions in the same bottle" architecture, but at least no one outside the bottle is hurt. In other cases it can look more like a well-refereed cricket match — polite, disciplined, and restrained, but just as serious to players and stakeholders alike.

The purpose of particularly important megasystem is controlling consumer costs through competition, rather than, as implied by the conventional model, letting uncontrolled monopolies "win" and set prices. This megasystem model uses antitrust laws to control what each competitor is permitted to do. A megasystem of superpowers has led, among other things, to hot lines to assist in mutual deterrence, confidence building, and mutual disarmament. In the private sector, a megasystem runs by laws against restraint of trade, unfair labor practices, and other government regulations. Judging from the reaction of various competitors, megasystem controls are a two-edged sword, necessary at times but dreaded, nonetheless.

A quite different, sometimes terrifying, megasystem is that of asymmetric combatants — the U.S. Army against the Viet Cong, the bands of terrorists (or freedom fighters, depending on one's viewpoint) and national security forces. That and other forms of competition, combat, and regulation are the subjects of Chapters 3 and 4. It is hardly necessary to say at this point, that these, too, determine what excellent companies can and cannot do.

The two sectors: the essence of Chapters 3 and 4

Although they are intimately connected, it is customary to divide the outside world into two sectors, the private and the public. The private sector, the marketplace and other commercial battlegrounds, is the subject of Chapter 3. Its focus is on competition, both symmetric and asymmetric, as a system of systems. Chapter 4 on the public sector focuses on and is limited to the "delicate balancing" of the business-friendly U.S. Government. Its contrast with the private sector is already great enough to show that, without extending the chapter scope any further, naively changing from one sector to another can be dangerous to one's health.

Organizations already in one of the two sectors may find the treatment of their sector as oversimplified and possibly biased, but they may be brought up short by how different the other sector is, even though it *too* is oversimplified to its members. The author is a proud alumnus of both sectors and should be forgiven by their veterans for being so gentle on them. Both can be vicious and abusive, of course. But even when as gently described as possible, each sector can snare the naive and unwary organization, no matter how excellent, if it tries to compete in a sector it has never entered before.

A public sector organization or professional may be appalled at being subject to tort liability and the general lack of dedication to a national cause after a change to the private sector. A private sector organization or professional may be appalled at the intrusive level of public scrutiny, not only by the government but by the media and special interest groups. And it will certainly be astonished that its honored tradition of cost shifting from project to project is flat out illegal in the public sector.

Notes

1. von Kármán, Theodore, a Hungarian-American aerodynamicist, professor, author, international diplomat, pioneer in aviation, and pathfinder in space, 1881–1963. A friend to many, including the author as a young man at Caltech.
2. The Continental was part of a currency issued in the 1780s by the U.S. Continental Congress to George Washington's soldiers and became worthless almost as soon as it was issued. Miraculously, it was accepted as a true debt by Alexander Hamilton as Secretary of the Treasury after 1789, restoring the U.S. status as a creditor and leading to the rapid growth of the new country.

chapter three

The marketplace and
other battlegrounds

*For every competitive system there will be at least
one countersystem.* [Re 91 189]

Competition as a system of systems

Competition in architectural terms can be defined as an attempt by one
system or organization to equal or surpass others to gain something of value.
Generally speaking, competition is regarded as fair and legitimate if the
opponents are well matched and the playing field is "level." A remarkable
amount of legal and political energy is expended in trying to make it so.
Trying to achieve a reasonable leveling is a central objective in such areas as
restraint of trade, international currency exchange, labor practices, and deterrence of war. Serious mismatches make for competitive instability, serious
damage to consumers, stakeholders, and innocent bystanders, and serious
military miscalculations. At its core is competition. Like economics and the
art of war, it is primarily about relative levels between the competitors'
capabilities rather than about their absolute values, sizable and important
as the latter may be. For example, there is almost as much concern over the
relative levels of nuclear weaponry between India and Pakistan as was the
case between the Soviet Union and the U.S. Widely different levels are more
dangerous, more escalatory, and more globally disruptive than their absolute
values might imply.

Thus, competition can properly be viewed as a system of competing
systems and, therefore, one that can bring the powerful tools of systems
architecting to bear on it. As examples:

The leverage, risk, danger, and opportunities are at the interfaces.
The efficient competitor looks for mismatches
and tries to exploit or eliminate them.

> *Both success and failure are in the eyes of the beholder —*
> and there are many beholders.

The last heuristic specifically recognizes the importance and influence of stakeholders in the competition in addition to players on the field, including the media, spectators, owners, friends, enemies, consultants, advertisers, government agencies, and more. There is a clear system purpose: a victory or defeat of the attempt to surpass others, the assessment of which depends on each stakeholder. Just ask the historian, or the Wall Street broker, or anyone who has won a battle but lost the war. There are rules of the game, whether supervised or unsupervised, stated or unstated, ethical or otherwise. Ask the fan of any spectator sport how the rules affect the excitement and pace of the game.

There are tactics and strategies by all sides. The participant without a game plan seldom wins or succeeds. There is an intelligence or espionage function. The participant who also knows, or accurately surmises, the other participants' plans and capabilities can attain a considerable competitive advantage, especially if the competitor does not have, or is denied having, comparable information.

None of the above basics are news to the experienced competitor nor to the excellent organization. Understanding the basics has been essential to their survival to date.

The system architectural perspective of competition as a system of systems is less understood and, consequently, the competitive advantage it offers is easily missed. The first step in its use is to recognize the nature of the battle; that is, whether it is or should be symmetric or asymmetric.

Symmetric competition

Symmetric competition[1] occurs when two or more sides have much the same capabilities, are playing the same game, and follow the same rules. Strictly speaking, it exists only over the long term and then largely as an abstract version of reality. In economics, for example, the law of supply and demand applies only on average over a protracted period; transients are frequent and external events, such as a war or a change in technology, which can upset even the best predictions.

Nonetheless, the concept of symmetric competition and of its rules provides a useful model of the real world in thinking about what organizations can and cannot do. Symmetric competition characterized the Cold War for decades. It was the basis of the calculus of mutual strategic deterrence — that neither the Soviet Union nor the U.S. could or would risk starting a World War III as long as mutually asssured destruction would result.

In the private sector, for all the talk of free trade and open competition, the loud demands of a "level playing field" — particularly by the present losers — have historically resulted in regulation, quotas, antitrust litagation, calls for changes in trade agreements, and other government intervention for centuries. Agricultural subsidies, government-set prices of food, housing, and other means of assuring survival of the disenfranchised were in place long before the industrial revolution in the nineteenth century. In the automotive industry and elsewhere, there were conscious decisions made by the largest two competitors not to drive a third competitor into bankruptcy, even if it meant higher costs to the consumers. When strong foreign competitors appear in a country, one of the first reactions of the threatened native companies is to request their government impose tariffs, inspections, quotas, "native content," and other obstacles to product entry. It is arguable whether these level playing field tactics are worthwhile in the long run. Suffice it to say that they are understandable under the urgent demands for protection and jobs.

If the conflict between profit and jobs is roughly balanced and controversy is unrestrained, then the situation will tend to be oscillatory (back and forth). Its dynamics can then be described metaphorically in organizational language using Newton's Laws of Motion expressed in organizational language:

An organization at rest will tend to stay at rest. (Inertia)
An organization in motion will tend to stay in motion —
up or down! (Momentum)
To every action there is an equal and opposite counterreaction.
(Force and energy)
With its counterpart,
For every management move there will be at least
one countermove in reaction.

The basis for such Newtonian organizational heuristics is not because "F = ma" but because organizational and physical dynamics resemble each other in reaction to external forces anecdotally if not in detail. But the Newtonian versions are easier to extrapolate. For example, response to change must take time. Once again from Newton: nothing can go from point A to point B instantaneously, nor once started can it be stopped instantaneously from going too far. To do either literally would require infinite force. Sound familiar? By analogy, small organizations can be energized more quickly than large ones, given the same applied force. Simple predictive concepts, these "Newtonians" are often forgotten, even by the best of organizations. We shall return to them later in another context. Bureaucracies are difficult to change because they are designed to maintain a status quo of interlocking rules.

This resistance to change, even of acknowledging the presence of competition, is well illustrated by the history of the manned spaceflight race.[2] In 1959, for example, strange as it may seem retrospectively, there was a serious, presidential-level debate over the question of whether the U.S. was or even should be in a race for space with the Soviet Union; this came only months after a series of Soviet spectaculars had laid down the challenge. It was yet another year before a recently inaugurated President Kennedy emphatically answered in the affirmative — 2 years after the Soviet Union had declared, de facto, that a race was already underway. The space race story, it should be noted, is by no means unique. Many an organization, thinking it is the king of the mountain, has ignored and even denigrated upstarts who dared challenge the king until too late. The lesson:

It only takes one to declare a race.

for which a demonstrably more effective response is to acknowledge the challenge and take whatever action seems best under the circumstances. It may even serve as a deterrent. Put on the track shoes, perhaps; get in shape, probably. Or, as Satchel Paige of baseball fame reportedly advised, "Keep looking back. Somebody may be gaining on you!"

Not surprisingly, once the race was started, it became difficult to stop. Although the Congress had authorized only the Apollo project, NASA kept going as if there had been a commitment to manned spaceflight indefinitely When the Soviet Union finally left the field of play in the late 1980s, the driving reason for manned spaceflight, a competitive race, vanished and was replaced by serious questions of the purposes of the space station and the Shuttle intended to support it.

A telling example of the countersystem heuristic in action is what might be called the life and death of systems analysis as a design methodology. In the early 1960s it was decided that the Department of Defense should establish, at the highest level, a civilian office of Systems Analysis staffed by the best and brightest, and devoted to analyzing military needs and finding the mathematically optimum way of meeting them. Their analysis covered weapon systems of all kinds, from aircraft, missiles, and tanks to submarines engaged in specific combat scenarios. New kinds of systems were conceived and built in an effort to achieve predictable military superiority in specific scenarios. By the end of the Vietnam War in the early 1970s, however, it became evident that there was an important corollary to:

For every competitive system there will be
at least one countersystem.
which stated that
A system optimized for a particular situation is unlikely
to encounter that situation.

The first heuristic implies similar sets of weapons fighting other similar sets. Tanks against tanks, aircraft against aircraft, ships against ships. The lesson learned in the second heuristic is that the opponent can make the same calculation ("my opponent wins and I lose, under the stated conditions") and conclude that there is no point in getting into such a fix, and thus to the extent that is practical, avoid presenting it. The North Vietnamese response to superior American firepower was a classic one. Orders were issued to never attack the U.S. head on. You will lose. The North Vietnamese strategy was shifted to the media, using small but very visible thrusts (for example, the first Tet "offensive") timed to coincide with or initiate U.S. Congressional debate.

In another example, the Soviet Union responded to high-tech American weaponry with massive numbers of tanks and aircraft. As a wry Allied expression went at the time, "We in NATO (about to be overwhelmed) certainly face a target-rich (lots to shoot at) environment!"

The real needs then were not for weapons optimized for victory in a matched-numbers scenario, but for larger numbers of weapons that worked reasonably well in a variety of scenarios, with excellent training in their use, and for first-rate warriors to take them into combat and prevail. Designs based on systems optimized to win in specific scenarios had failed. Designs with "good bones" and resilient architectures took their place. One example is the General Dynamics F-16 fighter aircraft designed for maximum maneuverability in visual air–air combat, that is dogfighting. Its actual combat use has been for air-to-ground support, a very different scenario but one in which maneuverability is a lifesaver. Along with other features, its electronic fly-by-wire control system greatly improved its survivability compared to the bulkier and, hence, larger cross-section, hydraulic system it replaced. It remains one of the most effective military aircraft in history and has become a fighter aircraft of choice in many nations confronted with a wide variety of military scenarios and opponents.

An even longer-lived architecture, this time in the private sector, was that of the DC-3, the first airliner to turn a profit for its owners. Built by the fledgling Douglas Aircraft company and designed by Arthur Raymond, it was a two-engine, propellor-driven monocoque, underwing, relatively comfortable, passenger plane that carried out missions of all sorts under all conditions all over the world for virtually every nation and company for more than 50 years, well into the jet age.

The commercial world knows this lesson as keeping a balanced mix of performance indicators, not just profit, market share, return on investment, or stock market P/E ratio, but all of them.[3]

Asymmetric competition

> *Sometimes the best way to defeat a threat*
> *is to do so "out of bounds."*

There are many other lessons learned through competition, of course. One of the most potent is to respond to competition in one area by competing in another, preferably one that endangers the foundations of the original competition. Military firepower can be countered by tactics of avoidance, stealth, and political action — even terrorism. In computers, superior but expensive operating systems for one customer group can be countered by ones easy for customers to modify, enhancing the customer base instead of competing with an existing one. Efficient software is countered by less expensive software calling for more memory to do the same job.

In the Congress, debates over principle can be converted to a surrogate discussion of procedures, leaving activists baffled and stranded. In the White House, a President can be faced with a single bill containing issues the President favors with others the President opposes, forcing a no-win choice of passage of an unwanted provision or veto of a desired one.

And then, using methods still not understood by the author, wives do seem particularly adept at asymmetric responses to issues raised by husbands. They stay on a subject when ahead and change the subject when not, seemingly effortlessly and without rancor. Husbands mostly huff and puff in frustration.

In organizational terms, any response to asymmetric or dissimilar competition must include early recognition of its real purpose and target. Attempts to respond head-to-head in areas of mutual strength such as financial clout, legal maneuvering, product monopoly, or market share may win a battle but lose the war. A more successful, asymmetric response might instead be against the competitors' logistics, product quality, treatment of labor, campaign financing, Congressional relationships, television docudramas, patent structure, recruiting of architects and executives, mergers with key suppliers, or other weak points of the competitions' defenses.

The classic analytic work on asymmetric competition, written in 1970 with the intent not to show bias to either side, is *Rebellion and Authority* by Leites and Wolf. [LE 70] Though somewhat dated, having been written in the middle of the Vietnam War, it makes instructive reading for excellent organizations who from one time to the next may find themselves to be on one side, say rebellion in the sense of a new product line or authority in an attempt to establish standards. It drew some interesting conclusions.

- There is no guarantee that an asymmetric attack will win, of course.
- It is a tool of the weaker competitor and wins only by exhausting the will or forcing a disproportionate response of the stronger.
- If overly vicious, any asymmetric attack or response may turn world opinion against the perpetrator, denying it outside support.
- Eventually a true competition or battle will be necessary or any perception of victory will prove to be hollow.

Two other specific responses, self assessment and business intelligence, are discussed in more length in Part IV, Chapter 9.

The origin and behavior of cycles and transients

To understand businesss and other economic cycles that affect the behavior of organizations, it is helpful to first understand the origin and behavior of cycles in general. Misunderstood or ignored, they can abort what might have seemed as fine opportunities because the timing somehow wasn't right.

Cycles, particularly fairly stable ones, are a well-understood response of so-called linear systems to input events or conditions. Based on the calculus of differential equations [GA 45], the mathematical model of linear systems has become so pervasive and so easy to use that many engineers and architects believe it to be almost universal. Indeed, for determining the effects of small perturbations on very complex manmade ("artificial") systems, linear mathematics does quite well. So well, in fact, that it has entered the general vocabulary in such words as transient behavior, cycling, oscillation, damping, stability, linearizing, and synchronization (in "sync"). Providing that the system is properly designed, it should be stable and predictable; that is, it shouldn't crash or self-destruct. Providing that outside interference stays within statistical bounds, the response of linear systems, even in heavy interference environments, also stays within statistical bounds.

Linear system transients tend to respond to events in two ways: continuing oscillations or one-time rebounds that peak, shudder somewhat, and then die out or stabilize at a new value. However, if the peaks and valleys of the responses to a number of past events happen to come together at just the right time, the sum can be very high — or very low — for many years. In business cycles, of which there are at least four, these sums can generate periods of prosperity or depression that can last for a decade or more even though none of the individual responses appear to be abnormal. Examples of events that generate cycles and rebounds are wars, previous depressions, abnormal weather, drastic currency devaluation, new technologies, and political upheaval.

It might seem that by careful study of all the smaller cycles, a large scale prosperity-to-depression-to-prosperity period could be predicted and even managed. The effects on a business of introducing new technologies at different times in a cycle might be estimated. Unfortunately, due to the number of cycles, the presence of random events, and interactions between the systems originating the cycles, neither prediction nor management are easily achieved — even when the relationships between *pairs* of systems are understood. Try as they might, economists have proved to be poor predictors. The rueful conclusion is that economics predicts wonderfully, but only in retrospect.

One of the reasons for this difficulty is that different events and their responses can interact with each other such that the resultant outputs are seldom replicable. For example, it is almost impossible to perform an experiment on a real organization, except for minor tweaking, that can be replicated as little as an hour later. Even then, it takes tight management to be sure that nothing different is going on in a subsequent attempt. We now realize that the real world has all the mathematical characteristics sufficient for what is now recognized as mathematically chaotic operation — a kind of erratic, unpredictable but contained, cycling that never quite repeats. Its theoretical genesis was aerodynamic turbulence but it is now known to occur in many other fields. Now that its nature is better understood, it is easily recognized in such phenomena as computer crashing, communication network lockups, weather prediction, and in otherwise impossible types of signal interference.

The cycles that most affect business opportunities are agricultural, industrial, national security, demographic, and electoral. Though they do interact with each other and may be difficult to disentangle, understanding them is certainly better than ignoring them and wondering why the timing wasn't right for the introduction of a new product line.

A century ago the main inventory cycle was agricultural, waves of over- and under-production of agricultural products by large numbers of small farmers. That cycling was considerably damped by offsetting governmental subsidies and by farming technologies developed by the Department of Agriculture over a period of more than 100 years. The result was an increase in agricultural productivity, a corresponding reduction in the labor force required, and a steady growth of agribusiness to the point where the monetary value of its agricultural exports, as a fraction of GNP, has remained essentially constant. The reduction in the size of the required labor force, however, did cause massive movements of the population from rural to urban areas, generating a new set of employment issues in its wake. Today less than 3% of the American population is involved in farming and its supporting industries compared to about 50% a century ago.

The industrial cycle is one of innovation, rapid growth, peak performance, over-competition, excess inventory, price erosion, and finally a commodity-like bottoming out. The cycle begins again when a new technology or idea gets a chance to start with labor and capital opportunities created by the bottoming out of an earlier technology. Depending on the technology, this cycle can take as little as 2 or 3 years, particularly in software applications, or up to 50 for the generation of energy or the production of new types of armaments.

There is concern today over whether the manufacturing sector, with its focus on productivity, is on its way down the same track as the agricultural one has been in which fewer and fewer people are needed to produce the same, or even more, manufacturing value. At the present time the latter is staying constant as a fraction of the GNP and the former has been dropping sharply (2 to 4% per year) for 4 decades. It appears that the next cycle will be in the knowledge, or smart machine, business, if it isn't already well underway.

Arguably, there is a national security, or nationalism, cycle as well, one which interacts with the industrial cycle from time to time. Such was the case in the 1930s when the 20 year, major war-to-major war cycle pulled the U.S. out of a 60-year Kondratieff[4] business depression. Equally controversial, and perhaps caused by the national security cycle, is the demographic or population cycle, one of alternate generations of population explosion and decline.

The effects of the U.S. electoral cycle, while relatively smaller and largely limited to the U.S., may be changing. During the long period of regional economies and a relatively isolated U.S., its 4-year presidential election process provided cyclic inputs to both the private and public sectors. The change to a global economy will probably increase the effects of the U.S. electoral cycle on other economies and, consequently, on international companies, heightening the incentive for them to be heard during the American election process, formally and informally.

Interactions with the outside world

There are a number of ways in which companies and agencies interact with the larger world. Four of the most important for purposes here are the economic viability, products and processes, sharing of added value, and the transfer of technology.

Economic viability

First in line must be economic viability. In other words, whether the venture makes any sense at all in global economic terms. To economists, four related questions properly answered in a mutually consistent fashion largely determine whether an enterprise is likely to succeed over the long run. If they can't be answered, something is wrong somewhere. The questions are

Who benefits? Who pays? Who supplies? and *Who loses?*

Answers are usually straightforward for commercial products in which the user benefits and pays, and the suppliers are organizations whose profit depends on user needs. Losers are those who do not believe that the costs are affordable within their budgets and priorities, but the choice to enter the market is theirs. There are a few implied conditions, however. International trade must be stable. The losers must be out of choice, not out of external pressure, or they may force major change.

More complex are public services. Until relatively recently, governments supplied them, the citizenry benefited and paid for them through a tax system. Generally speaking, laws made sure that access to the services was uniform in the cost to the individual. Every citizen was charged the same amount to send mail, to make local and long distance calls, to obtain weather reports and other publications, and to travel comparable distances. Over time, these services are being partially privatized and deregulated but, for the protection of the seriously disadvantaged consumer, partially re-regulated. In any case, access has been steadily decreased as it was shifted from being equal access to being cost-based; that is, accesses costly to provide have been reduced in the interest of decreasing costs and increasing profits. Major airline service to smaller towns, for example, has either ceased or become very expensive to the traveler. Because most passengers see lower costs, albeit less convenience, a shift in this trend is unlikely.

It is arguable whether the communication costs to the consumer have decreased; they demonstrably have become more highly taxed and are increasingly determined by fixed access charges instead of the amount of usage. To the surprise of many, the demands for a return to more regulation come not so much from the users as from the suppliers. Suppliers express their concern over unhindered competition, uncertain costs and liabilities, and, consequently, higher profit risk than anticipated. Consolidation, for reasons of economy of scale coupled with overlapping of expensive facilities (residential cabling and distribution), appears to presage a return to monopolies, but this time much less regulated and far more costly in marketing, sales, promotion, and legal costs.

Much of the difficulty comes from what is called "the billing problem." Price determination (billing) for a physical, easily quantified, product has been widely accepted for many years. Not so for a service. Service is a matter of social system priorities and politics. As Chapter 2 indicated, cost and value received are not the same. The former is a supplier figure, the latter is a user (observer) perception and it varies greatly with the observer, time, and circumstance. Consider how different observers perceive the cost, and cost distribution, of catastrophic health, message delays, total travel time, and such personal factors as age, state of health, type of business, and immediate financial status. Or consider access to weather services, navigation coordinates, space photography, and the like. When not needed, they are valueless. At other times they are so urgent as to be virtually priceless. They are desired when needed and not otherwise. So, how can a supplier bill for them customer by customer? It is so much simpler if the market can be aggregated and consolidated, as in a business or in a region. The latter, of course, means within the government or regulated by it.

Even discounting any personal biases by the author, it is clear that there are significant differences between selling a product and supplying a public service and between pricing according to user value and according to supplier cost.

This conclusion bears directly on the purpose of this book. Excellent organizations familiar with one should be aware that they might not do too well in the other. The downsides may be more widespread than simply financial. And the consumers and investors should be aware that if an organization is doing poorly, it is probably inefficient and, hence, more costly to them.

Products for a changing world

Assuming that a venture passes the economic viability test and has solved its billing problem, its next most important interaction with the outside world is in its sales and maintenance. Presuming that any excellent organization has mastered the necessary arts of marketing, advertising, distributing, and maintaining whatever it might sell — and is more than aware of environmental and liability concerns that could affect its processes — the essential question is one of the nature of the product in a time of rapidly changing technology.

Life would be, and was, much easier in a time of very gradual changes in technology and product types. Products could be treated as commodities — things that could be made by others — and competition meant striving to make them more efficient, less costly, and of greater value for the same or related purposes.

The high-tech world of today is different. High-tech suppliers dread the time when their products become commodities, when the only distinctions between their products and those of their competitors are in the color of paint, the style of the fenders, the colors on the display screen, or the glossiness of their brochures. In a high-tech world the objective is new kinds of products and applications which generate new kinds of markets, markets that didn't exist before and are rapidly becoming necessities, such as hand held calculators, direct broadcast television, high definition cable television, laptop computers, desktop publishing, Internet browsing, instant access to a world of information, global communications, drugs for a population that now lives longer, and so on.

Following close behind are new services — hospices for dignified care of the terminally ill, knowbots for connecting suppliers and customers all over the world, global paging, just-in-time inventory, computerized road maps and global positioning for the automobile driver, and the like. Services like these, previously unavailable, now are available to millions of people. The demand seems inexhaustible.

The question of the nature of a product to fit these and future markets is more than the design of a single one for today's market. It is a question of the architecture of a product line, a series of products over time which individually may last only a few years but which, as a line or family, may last for decades and for several changes in market composition and type.

As examples, the architecture of an automobile (an internal combustion engine, four wheels, tires, three to six passengers, power train, horn, headlights, bulk storage, etc.) had remained essentially the same for 50 years until the arrival of high-quality, microprocessor-equipped automobiles. Propellor-driven aircraft dominated airlines for about 40 years; jet aircraft architectures have and will for at least that long. Submarine design, aircraft carrier design, armored tank design, residential design, and others are similarly long-lived. Even computer architecture (memory, processor, display, digital logic, separate operating system, applications software) has been much the same for 20 years, though the rate of change of technology and the size of some components is more than 10 times as fast.

The value of an architecture, therefore, is maintained considerably longer than the value and life cycle of a single product. If it has "good bones" like successfully evolving mammals have, it can survive for extraordinary periods and adapt to a wide variety of external environments.

But the risks are high in choosing the right approach. As Bob Spinrad of Xerox explains, "You better get it right or you will live with your mistakes for 20 years!" He should know what it means to make an architectural change in an organization. He was deeply involved just a few years ago in changing Xerox from an excellent copier manufacturer to "The Document Company."

In a talk to a USC architecting class Spinrad provided the following examples of difficult, even dangerous, issues in the realm of product architectures. The specifics are less important than their diverse nature. No question has a single correct answer. All affect multiple features of each product in a product line.

- In a line or family of increasingly higher performance products, which one is the design point (or center) that defines a standard architecture? Is it the middle one, with deletions below and options above? The top one with deletions? The bottom one with the fewest standard features? How is upward and downward compatibility assured so that customers can migrate upward as their fortunes improve?
- How can the products be differentiated from each other and from those of the competition and yet hold or agree to uniform standards for all competitors?
- Options are fine but how can the creep and drift of requirements be constrained?
- How can customers migrate to or from this company's products to another's? Should they have the option or should they be locked into ours?
- How can an architecture be open or modified and yet owned and controlled? [MO 93]
- When is the best time to abandon one architecture for a new one? Should the new one be backward compatible? What is forcing the change?

To complicate the design still further, the answers must hold for at least 10 and preferably 20 years. For some essential products and services such stability may be possible, but not necessarily for the newest markets of Internet, global positioning, laser weapon systems, building materials, consulting services, environmental control, long distance transportation, and space launch vehicles. The architect's problem is to make the answers valid for as long as possible and to know when to abandon that architecture for a better one. The Introduction to Part III in its story of the six eagles gives a particularly dramatic example of how difficult it can be to decide on abandonment.

A directly related problem is the architecture of the organization that creates and builds the architected product or service. It is often said that the architecture of the organization should match that of the product. That is, its units should be similarly aggregated and partitioned. For a single product line supplied by a single supplier, that seems obvious. But what of an organization that supplies a number of quite different products? Chapter 9 of Part IV picks up this question, so it will not be answered here. But think about it in the meantime, if only to appreciate the implications of different answers to the same question. There are no "right" or "optimum" ones, incidentally.

Sharing of added value

A third interface with the outside world is the exchange of added value. A company creates a new product, adding value, which it sells to an outside customer, keeping part of it as profit and experience. Straightforward.

Now consider the value added by this company to a component supplied by another company, making that component more valuable. Perhaps new application software is added to a microprocessor, potentially increasing the sales of the microprocessor to suppliers of other markets, but only if that software were available to them. Perhaps a far better energy control system for an electric battery was added that made it competitive for electric cars. Perhaps there was the addition of software smarts to instruments enormously increasing the value of these instruments to more customers. Should the company be the sole benefactor of the value it alone added?

Clearly comparable in importance to the creation of value is the distribution of it. It is not enough to create value for a customer, shareholder, and user. The creator must share some part of it or create no more. Should the added value be further shared?

An example from the hard-headed arena of real estate will help illustrate the general principle of value sharing. At first look, the result will seem to be quantitative, an easy problem for an economist. But even in this quintessential example, that could not be the case. Something called "fairness" had to be brought into play.

Realtors are particularly aware of the necessity of distributing value added in a transaction. As an example from this author's personal experience, Company A wished to purchase a piece of unused property from Company B. The property itself was a corner piece of a city block, the rest of the block already owned by Company A. Of itself, to some other buyer or to be sold at a price comparable to others in the vicinity, it had been fairly appraised at $10 million. Its potential value for some other usage could be as high as $30 million, but whatever was built on it would then have to cost at least another $30 million to make economic sense of an asked $20 million investment in the land under it, something unlikely in the immediate future.

To Company A, which owned the rest of the block, it made possible an economic development of the whole block without building an otherwise required parking structure at a cost of $3 million. Also, for reasons of security, Company A did not wish to have a high-rise neighbor that could look down from above and peer through windows into classified areas inside. But should the property be developed as a high-rise facility by another buyer, it would require Company A to remodel its buildings to protect the security of its work.

Company A offered $10 million, the appraised value for that property in that neighborhood if sold to any other buyer. But Company B responded by asking $20 million based on potential use. The reasoning of each party seemed unassailable, or so it seemed to this author.

The realtor suggested $15 million, not because it was half-way in between but because it was fair to both parties. For Company A's purposes, the additional price was offset by cost avoidances, a fact well appreciated by Company B which happened to know Company A very well through business relationships. For Company B's purpose and to show its top management that the property had been sold at a considerable profit, it was gaining $5 million above the appraised price and giving up only a rather unlikely future opportunity for more. The Boards of Directors and the government accepted this reasoning and the deal was closed. At $15 million everyone gained. At any other figure, no deal, and everyone would lose.

In this case, the added value was created by Company A, but it was only realized by sharing some of it with its "supplier," Company B, which had the right to refuse to sell it, or worse yet, the right to sell it to someone else. Interestingly, that obstacle was overcome by Company B realizing the importance of Company A's high-security business to the nation and with good grace concluding that Company A was probably the "best" buyer in a larger sense. Value, clearly, has many parts.

In a high-tech world, refusal to sell or license a product or service, usually for proprietary reasons, has had mixed consequences. Morris and Ferguson [MO 93] would argue that refusal to license its operating system cost Apple dearly in its computer hardware business. In contrast, the U.S. Patent Office would claim that ownership was a right that underpinned the whole concept of patents and copyrights.

A variation on refusal to sell is to offer to sell only at a price which so reduces the customer's estimated profit that the customer elects not to proceed even with development. The product is then not produced, the market is never sampled to determine the final price, and everyone including the final customer loses. Examples can be found in medical diagnostics, in aircraft maintenance systems, and in computers and software where good ideas of uncertain value but commendable applications in unknown markets were never ventured.

It is to the credit of the U.S. Defense Department that it found a way to, at least, explore radical ideas of undetermined value to the point of demonstration. It established a special agency, the Advance Projects Research Agency (ARPA, now DARPA) for the purpose. Some of the anticipated results occurred as predicted. About half of the ideas proved unworkable but valuable nonetheless for reasons discovered in the process, thus adding to the experience base. At least another third proved not only possible but very valuable, but not necessarily for the reasons given in the beginning. Surveillance spacecraft, computer sharing, ARPANET (the predecessor of Internet), the initial Deep Space Network, stealth aircraft, and many others escaped the deadly trap of premature assessment of value through the ARPA mechanism. Instead, they reached the nirvana of a completely unpredicted "latent" market of huge size.

Some highly successful systems reached their own nirvana through sheer luck. In a fine example of the "law of unexpected consequences," in its beginnings the now ubiquitous global positioning system (GPS) "breakeven" customer base was estimated as at least 20,000 users taking a few navigation fixes per day, a figure deemed so improbable that only a critical defense need (submarine navigation) kept it alive. The actual, but then latent, market for GPS turned out to be millions of users from forest rangers fighting fires to geologists looking for oil and boat skippers searching for their harbors in dense fog. Wise decisions by the U.S. government made the added value of the GPS available to the world at only the marginal cost of receivers, instead of trying to recover the costs to the Department of Defense for its development, supply, maintenance, and operation of the quite costly satellite constellations.

More examples are easily found, some in the private sector. Bill Hewlett of Hewlett-Packard put hand-held calculators on the market with results remarkably similar to those of ARPA: modest expectations in a limited market that exploded into huge returns in a larger one.

Research and development support by the outside world

Unfortunately, sharing added value likely to benefit a *competitor* is, perhaps, a major reason that for-profit organizations sponsor very little basic research. When they do, it is highly protected by them, even to literally locking it up and throwing away the key in order that it not destroy their present markets. For the added value to be fairly shared, it seems necessary to sponsor it sufficiently broadly that it benefits someone, unknown at first, within the support base. For some research projects that criterion could require a base as large as the population of a nation; that is, the project would require the sponsorship of the Federal Government and its agencies like ARPA and the National Science Foundation (NSF).

Sharing added value is an important factor in determining what an excellent organization can do. Directly supporting basic research may be asking too much, but knowing and supporting its general value is not. Tracking research efforts carefully in technical areas close to those of the organization can suggest ways of moving in new technical directions; ignorance of them precludes those options.

In any case, sharing added value with both customers and suppliers seems to pay off not only in improved relationships but as a way of creating new markets.

Summary

The opportunities and constraints on what excellent organizations can do depend upon the outside world within which they must live and work. This chapter showed that the outside world is one of systems, of symmetric and asymmetric competition, and of a variety of interacting cycles and responses to events. It indicated the importance of long-lived architectures in a continually changing world and of sharing with others the value added to the end applications.

Notes

1. For enjoyable and instructive reading on symmetric competition, see two USC SAE Research Reports: "Utilization of Three Design Theories in a Real World System (the architecture of a professional football organization)" by Jim Cracraft, 1989, and "Heuristics for Playing Board Games" by Thomas L. McKendree, 1988.

2. The reader may perceive that a disproportionate fraction of examples, good and bad, are in the governmental sector, especially from NASA and the Department of Defense. The reason is less the author's close association with both than because these agencies are held up to public scrutiny and are, properly, on stage all the time. Hence, well-documented, widely known, and very public examples are more numerous (and objective?) than private ones.

3. One of David Packard's (of Hewlett-Packard) best rules for success was, "We price to make a profit on the first item." It worked because H-P's high-tech product strategy was: in first, high profit, on to the next. Catch us if you can! It was counter to prevailing market share strategy of start below cost, grab market share, and rake in the profits over the long run. Few but Packard realized that market share is a commodities model, not a high tech one.

4. One such prosperity–depression–prosperity cycle is the Kondratieff Cycle, named after Nikolai Kondratieff, a Russian twentieth century economist, who postulated a cycle time of about 60 years based on the supply–demand cycle of commodities and energy sources. [HAW 83 66-73] Joseph Schumpeter, an Austro-American, an early twentieth century economist modified the causes, thereof, to technological causes and showed correlations with periods of war. The cycles, though large in scale, did not necessarily include all countries, e.g., Great Britain did not experience the severity of the 1930-40 depression in the U.S. and elsewhere. Peter Drucker, an American economist, has written extensively on the subject [DR 81], as have Paul Hawkin [HAW 83] and Jay Forrester [FO 82 a/b]. Forrester's work gives some credence to the "long wave" theory by adding more recent technological information and confirming computer modeling. If true, however, the predicted depression (1930–40 + 60 yrs. = 1990–2000) presumably is either late or been diminished by the impact of information technology — a factor ignored until recently by most economics theorists.

chapter four

"Delicate balancing" in and by the U.S. government

Introduction

Governments play many roles in responding to national needs. Four of the roles affect the way in which organizations can respond to global change. Governments provide national security, without which the citizenry and its institutions could not survive. Governments regulate the national economy and participate as a buyer and client in it. And they assist their citizens in foreign trade by establishing mutually agreed upon arrangements with other governments.

It is well beyond the scope of this book to cover even these four roles in detail, much less to include their effects in other countries. This chapter, therefore, is limited to how these roles are carried out by the U.S. Government and its effect on excellent organizations faced with global change.

The role of elected governments in human affairs

An elected government, certainly that of the U.S., is, as Abraham Lincoln so elequently phrased it, "of the people, by the people, and for the people." It is *of* the people because its members come from the people and are directly responsible to them and to no other authority: religious, civil, secular, lay, or foreign.

An elected government is *by* the people because it is by their authority — and only their authority — that a government exists. No king or military force established it or granted it rights.

It is *for* the people because the Constitution so limits the government by internal checks and balances that it has very little opportunity to be otherwise. The framers of the Constitution from their own experiences and history had ample reason to be wary of government, any government, but at the same time they were well aware of its necessity. Given the rare opportunity in human history to construct their own government, they acknowledged the need for national defense and the public welfare, but devoted most of the Constitution to *restraining* the government, reserving everything else to the citizenry.

It is difficult to overestimate, but vital to understand, the impact of this restrictive approach to government, not only on the behavior of the government itself, but on the nature of what excellent companies can do under its jurisdiction. On the one hand, it offers them protection, services, and freedom. On the other, it regulates and constrains them (on request, it should be pointed out) whenever the citizen's perceptions of national interests so dictate.

Maintaining the "delicate balance" between freedom and control against the upsetting forces of human societies is the government's most difficult, and valued, responsibility. It has fought wars to establish and maintain that balance. Although its deliberate checks and balances of authority have been the frustration of activists, entrepreneurs, civil servants, presidents, and foreign states in their individual times of need, they are most unlikely to be changed. Historically, only conquest, corruption, anarchy, or ideological partisan warfare have been able to upset duly elected governments.

On maintaining the delicate balance in and by the government

There are both positive and negative consequences to purposeful restraints on an elected government. The positive consequences mean great freedom for excellent organizations to act as they think best. This freedom has meant an extraordinarily favorable atmosphere for innovation and societal change. The principal negative consequences are governmental delay and uncertainty in decision making, and the unpredictability of its future commitments.

Almost any action by one branch of the government can be nullified by another, or even by itself at a later time. Thus, no one person or branch is the final authority and no single branch can make an irrevocable commitment to much of anything, national or foreign. In effect, all branches must approve, which takes time, understanding, and consensus. The citizenry would have it no other way.

The resultant consensus makes for an enormously powerful country once energized or provoked, but the accompanying delay in reaching it can generate a pendulum-like oscillation that can wreak damage on those caught by the swings in policy. Heading the list of those damaged can be large, excellent organizations caught going the wrong way at the wrong time.

There is indeed a reasonably documented argument that governments can make difficult situations worse by finally passing a law or taking action just as the situation resolves itself, creating an equally difficult over reaction in the other direction. Stimuli for job creation are activated just as the country faces a worker shortage. Production of new weapons is begun just as the threat that generated them fades away. Governmental affirmative action is dismantled just as its dilemma of help-without-preference is resolved and the program is continued in the private sector.

Deregulation is accelerated just as those who demanded it now demand that it be stopped and regulation increased. MCI demands that AT&T be directed to raise its prices so MCI can survive. Airline customers demand that major airlines be forced to continue to serve unprofitable routes. New electric power suppliers are demanding guaranteed prices now that the large public utilities (their debt burden for older and less-efficient power plants transferred to government-backed bonds) can build much more efficient plant themselves. Telephone users, used to paying only for phone calls they made, now find themselves paying for access to telephone facilities whether they use them or not — a major cost shift from businesses to residences. As predictable as a pendulum, deregulation will swing too far followed by regulation that swings back too far, and so forth and so on., et seq.

An architect–engineer would describe all of these phenomena as an oscillation of a quasi-stable system — like a pendulum, perhaps — needing a delicate balancing to keep it from destroying itself. In Newtonian-like insights:

- The acceleration to the left is maximum just as swing reaches its highest value to the right.
- The velocity is greatest in either direction while passing the middle ground.
- Most of the time in a cycle is spent at the extremes.
- The swings will last indefinitely unless "damped" by restraint or diminution of the driving force, whether gravity or special but countering interests.

Some years ago, a national news commentator, Fred Friendly of CBS, presented an educational television series entitled "The Delicate Balance," in which some of the most difficult issues facing government — such as states rights, the death penalty, abortion, and monopolistic business practices — were presented in quasifictional cases to groups of individuals that had been deeply involved in them: judges, lawyers, doctors, generals, activists, minorities — the lot. The moderator increased the specificity and difficulty of the issue as the situation evolved. Whatever the viewer's preconceived feelings were before the broadcast, it was almost certain that they would be modified, even reversed, by its end. Nothing, absolutely nothing, was as simple as initially imagined. The end of each program, incidentally, never resolved the issue. It left the viewer up in the air, the participants fuming at, but respectful of, each other, and the moderator exhausted. The viewer usually left with a much deeper understanding of why a "delicate balance" had to characterize whatever the resolution might turn out to be.

Rules count

*The more rules there are, the tighter the constraints,
and vice versa.* (Paton, Eric 92)

Before taking up the roles of the government in more detail, it is important to understand that elected governments operate by a system of rules (laws, regulations, directives, established precedents, and so on) to which they adhere with great tenacity. In a government, individuals succeed by following those rules to the letter. They fail if they break them.

But as citizens in a free society, individuals succeed by tolerance of others. When rules do exist, that means applying them to fit the occasion. This fundamental difference between tolerance and rules is a built-in cause of tension between the private and public sectors. It is a tension not easily changed, again because the public would not have it otherwise. They want the government and everyone in it to stick to the rules which they, the public, established. Sticking to the rules means predictability, even if annoying, and stability, even if somewhat oscillatory.

Metaphorically, the public wants a smooth ride in as calm a sea as possible. They do not want a government that keeps rocking the boat. But, once on the boat, the publc wants as much freedom to move around as safety permits.

Principles (like freedom, justice, equality for all, stability, and human rights) and management approaches (like efficiency, cost control, free markets, shared value, fairness, and level playing fields) unquestionably are important objectives. It is not coincidence that many of former are in the Preamble to the Constitution. But when it comes to making decisions, explicit rules are what really count. Without rules to define them, principles are meaningless. American judges instruct juries and litigants alike on the differences between rules ("the law") and personal opinions of what constitutes justice, "This is a country of laws, not men." A duly legislated, signed, adjudicated, and enforced law may or not seem to all parties to be fair, but it is explicit and not amenable to the whims of kings, presidents, dictators, religous authorities, judges, or juries.

This is not to say that rules can't be abused. They can. They can be misapplied to circumstances unforeseen or unintended by the rule makers. Rules often are made based on particularly egregious incidents — monopolistic practices, blatant misrepresentation, false claims, harmful products known to be harmful by their makers, harassment, and so on. The worse the incident, the more severe the rule. As lawyers know:

Bad cases make for bad law.

because other parties in other circumstances find themselves inadvertently vulnerable to its poorly written provisions. Tort liability, antitrust provisions, and cost-shifting regulations are some of the more notorious examples. Organizations contemplating change should do their best to avoid such legal quagmires whenever possible.

For example, suppose a monopoly knowingly and viciously gouges the public and drives competitors out of business. The victims petition the government for aid. Antitrust and restraint of trade laws are then made to criminalize these practices. Somewhat later, a literal reading of its provisions seems to permit the laws' application to entrepreneurial ventures having spectacular success with a new product that pushes all others aside. Even if the ventures are acquitted of violating the law, the publicity and costs drive the company into bankruptcy, discouraging others from innovating and exploiting their product lines. Unintended, certainly. Unexpected, no.

Rules about rules

Whether made by the government or by companies, rules almost always require interpretation if they are reasonably to fit the circumstances and maintain the "delicate balance" between asking for too much and doing too little. Formulating rules that are fair, consistent, practical, flexible, and consistent is not easy. Ask lawyers, legal staffs, policy writers, regulators, and systems acquisition specialists what it takes to draft acceptable and enforceable rules!

Knowing full well how rules are made, good rule makers duly write in enough checks and balances, waivers, and exceptions to block the worst proposals and pass the best ones Some of the rule makers' own rules on rules are

State the principle underlying the rule first.
Describe the context, precedents, and rationale next.
Make the rule as simple and understandable as possible.
Balance competing interests but don't harm the good
in deterring the bad.
Always leave an escape hatch to handle the unforeseen.

Living with the rules of government

The author was quietly apprised of how effective rules are made by a veteran of the Pentagon, a professional administrator who knew virtually every rule in the book and no doubt had written some of them. When his new and enthusiastic Director asked how to get things done, given the overwhelming number of rules enveloping this particular directorate, the veteran's reply was, "If you believe what you want to get done is right, there's always a way to do it. If it is illegal, there is almost no way without going to jail. Tell me what you believe is needed. If it's legal, I can always find a way to do it."

He went on with a gentle admonishment to his Director, "Assuming that a rule does fit, does represent an accepted compromise, and is balanced,

Don't tip the balance of a reasonable rule. It may fall over on you!"

The insight is by no means unique to the public sector. As seen in Chapter 2, it applied with a vengence to the Big Three automakers when they tried to upset the rules on automotive trade with Japan. It applied to unions when their pay demands forced the closure of noncompetitive operations and union jobs simply disappeared.

Architecturally speaking, stability and balance in the rules are in almost everyone's interest. It clearly is counterproductive to shout at a court judge, to impugn a referee's motivations, or to threaten to quit the game in an attempt to change the rules. Rules are only changed by general need, by general consensus and negotiation.

The down side of a body of rules is the nature of rules themselves — there never seem to be enough of them to answer all the questions that come up as the original rules are put into practice. Each new problem, each new use, each new design seems to call for amending, revising, or waiving an existing rule. On the other hand, eliminating them can destroy the "corporate memory" from which they were derived.

The government is well aware of that lesson, too. Several administrations and departments have tried waiving whole volumes of contractual specifications in the hope of saving money. One such was for a major aircraft buy. The result was that the winning contractor voluntarily applied almost all of them, anyway. There were no better ones; they did indeed represent best practices and they were understood by all parties. The principal difference was that the government was no longer required to monitor them. The contractor and subcontractors did monitor them in their own interest.

Burdensome as rules might be, in them lie the constraints and opportunities for those who must deal with the government. Know the rules and know them well — or find that veteran administrator who does. Make them into assets that help get things done, and done well.

To the organization contemplating the radical change of entering or leaving the governmental arena as a public servant, a public service supplier, or a government regulated entity:

> *Don't expect the government to be run like a business.*
> *Don't expect a business to be run like the government.*
> *They* can't! — and for fundamental reasons.
> And, above all,
> *Hard as it may be, it is easier to change yourself*
> *than either of them!*

It has long been astonishing to this author that the first insight is rejected by those in the private sector, the second is rejected by those in the public sector, and the last rejected by both as they each battle to change the other. The first group argues, with good reason, that the government should be more business-like. The second, with equal intensity, decries the lack of accountability and dedication to the public welfare of "contractors." At the core, the arguments are all about the responsibilities of each as (mis)perceived by the other. One of the most sensible conclusions to this argument was reached by a National Research Council report which said, "...the responsibilities are determined, in the end, only by the mutual agreement of all parties to the [common goal]." [NA 94, pp. *x* and 67]

Rules in a period of dramatic change

One of the consequences of rapid global and technology changes is that the valued old rules *must* be changed in response. Some of these changes can embarassingly change one's position on a number of hard-fought rules from the past. Suddenly, good rules for yesterday's world or technology become obstacles for today's. To give just a few examples of position-reversing questions in this changing world:

- Should Chrysler be entitled to U.S. benefits denied foreign automotive companies now that it is owned by a German firm?
- Should the restrictions imposed on Honda now be imposed on Chrysler or removed from Honda's U.S. "transplants?"
- If a U.S. telecommunications firm is bought by a foreign firm, should national security prohibitions on reliance on foreign parts for defense-critical telecommunications links then be applied to that firm?
- Has deregulation of the Bell Telephone System accomplished what was intended, given the recoalescing of all telecommunications to each home as cable operators offer phone service, as regional telephone services begin competing for long distance services, and as Internet connectivity becomes universal?
- In other words, can deregulation overcome the inherent properties and emerging values of systems?

The answers, themselves, are not the province of this book. Providing them is up to our oft-maligned, sometimes confused, Representatives in Congress. Rather, the point to be made here is that the effect of the same rule under different circumstances can easily be different. It can even reverse the positions, constraints, and opportunities many organizations previously assumed were permanent. The more narrowly the rule is written, the narrower its application and the more careful the user has to be in its application elsewhere. Which leads to the conclusion of this section and the underlying rationale for the rest:

> *When dealing with a government, respect the rules,*
> *the rulemaker, and the rulekeeper,*
> *particularly in a period of dramatic change.*

because therein lies the opportunity. To understand both the rules and the driving force behind the change (very often the government) can create opportunity where none seemed to exist before.

The government as a provider of national defense

The single most important balance in government, one that directly affects its survival, is that between the defense forces and the civilian sector it is supposed to protect. The citizenry of the American Colonies, having watched British forces protect, regulate, control, judge, and shoot them, knew what an imbalance could mean.

The countries in Latin and South America, all through the Mediterranean, Africa, the Middle East, and Asia, both before and since, know this just as well — and have had to re-learn it just as often. The defense forces in most of the countries of the world have all the weaponry, all the troops, sufficient justification, and all the dedicated believers to take over the country whenever they believe it to be in their country's best interests. They typically are efficient, well organized, cost-effective, and disciplined. In many countries the armed forces are a virtually autonomous part of the government. It has taken very little to go wrong in these countries for the proverbial "knight on horseback" to come riding over the hill to the rescue, so to speak, and take over the governmemt by acclimation.

Such an image, thankfully, has probably never entered the minds of most modern-day Americans. Impossible! Unthinkable! Thus, although the Constitution tasks the government to defend its citizenry, it does so with strong restraints on and control of the military forces, of how they are commanded, and how they are supported. In early 1969, Robert Moot, then Comptroller of the Department of Defense, expressed it this way:

> "The Constitution and the Congress, in their wisdom,
> made sure that
> the Defense Department could not work too well."

This verbatim quotation so impressed, consoled, and guided the author that he has never forgotten it. A friendly, understanding, and experienced man, "Bob" Moot preceded this statement by something like, "You are all new here, full of dedication and ideas of how to better defend the U.S. You may be frustrated and concerned that they do not happen in what almost anyone would consider a reasonable time. It is not the Department, nor the President, nor individual members or staff of the Congress that causes this state of affairs. It is because...and so on into the quotation.

Even so, the restraints can be onerous and do lead to tensions which excellent organizations interested in Department of Defense business must understand if they are to be both successful and helpful. At times the tensions can become quite personal.

For Constitutional reasons and often deep personal beliefs, U.S. military personnel are insistent and vocal that they are under civilian control, almost routinely acknowledge that control, and have sworn allegiance to that principle and the Constitution that established it. Parenthetically, this deep set of beliefs, a strength to be welcomed in any military officer, can be perceived as a weakness in the less disciplined, more ambiguous, free-wheeling world of competitive free enterprise. There is a second and arguably equally important aspect to the military–civilian tension — the difference in reaction time to external threat. The military commanders are confronted with situations that demand quicker action than before. Wars, as they point out in all sincerity, can be won in 6 days and lost in 6 years. Yet, civilian authorities, for equally serious reasons, increasingly believe in reaching regional if not international consensus first, even if it takes months or years to achieve.

Another consequence of having a strongly restrained, civilian-controlled military establishment is that the Congress and President can — and do — use the military establishment for purposes other than military, complicating the primary task of achieving an efficient, effective military force, but for what are politically important reasons. The exemplar of such purposes was the 1945 desegregation of the armed forces by President Truman decades before the civilian sector followed. He ordered. They saluted and obeyed. Only later was it seen as a good, practical, and honorable idea.

When the military–civilian partnership is well managed, major accomplishments are possible. The author can think of no better recent example than the recent, orderly, mutually supporting downsizing of the military, civilian, and industrial organizations. That absolutely necessary, massive restructuring was a management *tour de force* with far less turmoil, under more severe strain, and with many more positive results than have been achieved, say, in the American automotive and telecommunications industries.

What is the lesson to be learned by excellent companies? The Defense Department is only as efficient, strong, and protective as we, in our own best interests, have chosen to make it. It does not make the rules, it obeys them. It does not declare war, it wages it. Its job is a difficult one requiring understanding of the inherent tensions among its military, civilian, and industrial components and their disparate interests. A delicate balance is needed among their disparate interests. Excellent organizations should help maintain that balance in their *own* interests. In short, all would do well to learn the language, principles, and, yes, the cultures and imperatives of the others if, as a partnership, we are to be well defended.

The government as a regulator

Regulate: To control or direct in agreement with a rule.[1] The regulators: The Federal Reserve Bank, the Federal Trade Commission, the Interstate Commerce Commission, the Environmental Protection Agency, The Federal Communications Commission, the Atomic Energy Commission, OSHA, NRA, etc., etc.

Of all the services governments provide, regulation is probably the most contentious. Yet, regulation and control are *assigned* tasks; they weren't chosen. They were tasked by the public of its government in its own best interest. The risk to new entrants in a regulated field is the oscillatory behavior from regulation to deregulation and back again If that government strays very far from the delicate balance of too much or too little regulating, it hears from the most affected interests loud and clear, which may increase the extent of the swings to the damage of all parties concerned.

The public knows that any organization or system, without rules to control, protect, and nurture it, cannot last long. Without rules there is anarchy. Without rules there is no "free trade," and no stakeholder group knows this better than the traders. Competition requires a referee, an ombudsman, a "controller" who can enforce the rules in the common interest; someone to keep the playing field level when economic conditions would overturn it; someone to cushion the dynamic swings of the free market; someone to keep order on the streets and freeways; someone to guarantee service to all the citizenry, not just the most profitable sector.

Regulation, like rules, is a balancing act. As more than one executive has remarked, "I know I'm right because I'm being attacked from all sides!" Another was fond of saying, "My job is to distribute the pain. If everyone feels the same pain, no single one can object without being shouted down by the rest, with examples of their own!"

Although the differing perception of pain may be difficult to determine or balance, few people would disagree that whoever inflicted the pain should pay the cost of remedying it. Often such payment is made voluntarily, occasionally it takes a court suit to guarantee it. The contention lies elsewhere — in the legal quagmire of tort liability in which the designers and makers of a product can be ordered to pay compensation well beyond damages incurred during unintended, unpredictable, or abusive use. Organizations long in the private sector are well aware of the risks involved. Those most likely to be caught unaware are organizations or individuals accustomed to working for governmental agencies that elect to move into the private sector. Enough said about a serious problem for excellent organizations facing such a change.

The government as a buyer and client

The best place to start in examining the role of the government as a buyer and client is with the Congress; that is, the part of our government that budgets for and pays the bills for the needs of the American public — roadways, subways, military equipment, airports, public service communications, weather satellites, scientific research, and social services, for example. The Congress determines what will be supported, what will be acquired, how and by whom it will be funded, and the manner in which the Executive Branch will follow its instructions.

Ideally, perhaps, the Congress would act like a Board of Directors directing its Chief Executive Officer (CEO) and staff; that is, the President and the

Executive Branch. In practice, and because the amount of money that flows through it is in the trillions of dollars and because the ways in which it flows can mean prosperity or poverty to a member's district or state, the Congress often delves into the details of individual projects, just like citizens do in making their purchases and obtaining the services they need. By budgeting, authorizing, and appropriating, it decides whether the country should go to war and/or to the moon (it did both at the same time), go to the support of another nation, or to underwrite much of the world's scientific and technological future.

From a systems architecting perspective, it is the right and responsibility of the Congress to make value judgments in the public interest; that is, to decide what is good, bad, affordable, or worth less than some other need.

The political process and system design

The political processes of the Congress can have and has had significant impact on system and product design [FO 97]. The Congress, for example, alternately directed that the Shuttle launch vehicle not carry, then carry, and then not carry a potentially dangerous, hydrogen–oxygen-filled Centaur upper stage into low orbit, depending on Congress's perception of risk to the astronauts and the needs of missions of most interest to the members. Did the Congress have the right to do so? Of course. The right rationale? Well, that depends on what kind or rules one lives by. By *its* rules, yes.

The political process, the heart and soul of a democracy, operates according to rules established for it and developed by it through long experience at balancing values. Dr. Brenda Forman, a close observer of the political process, has codified and provided examples for a short list of those rules that particularly affect the government's relationships with high-tech product organizations [FO 93 and FO 97]:

Fact of Life No. 1
*Politics, not technology, sets the limits
of what technology is allowed to achieve.*

Fact of Life No. 2
Cost rules.

Fact of Life No. 3
A strong, coherent constituency is essential.

Fact of Life No. 4
Technical problems become political problems.

Fact of Life No. 5
*The best engineering solutions are not necessarily
the best political solutions.*

Forman concludes in an honest and straightforward way, "It (the political process) is not incomprehensible, it is merely different. Being understandable means being usable. For example, Fact of Life No. 3 is the route by which poor rules can be changed by those negatively affected."

To these rules might be added two that directly affect project management and the constraints and opportunities for excellent organizations:

1. *Congress operates based on annual cash flow,* not on multiyear life cycle costs, total costs, or performance — regardless of presumed economic or programatic benefits.
2. *Cost overruns will be fought with real ferocity* and not only by the Congress.

Other stakeholders in the national accounts will quickly join in the fray. Valid schedule delays and performance deficits may be accepted conditionally, but cost overruns cause real trouble. Any suspicion that these deficiencies could have been overcome by other means leads to disciplinary funding *cuts,* not supplements. There is a very practical reason for this imperative. Once budgeting is complete and appropriations (funds) are established, the unplanned arrival of an overrun, even a perfectly valid one, means taking funds away from another effort that presumably was doing well. The incentives to the latter to continue to do well decrease, to overrun increase, and things go from bad to worse. The message is all too clear — spend or commit irrevocably every annual allocation as soon in the fiscal year as possible. Never rely on carryovers of funds from previous years, they will be swept up and given to someone else. Needless to say, funded overruns in one project make unforgiving enemies in others. For example the most dedicated opponents of NASA's space station are within NASA itself — the science projects cut by a factor of three to accommodate space station misestimations and overruns.

Tellingly, these heuristics and characteristics are better understood on the political side than on the engineering side. Excellent companies would do well to remember them. Too few do and attempt to do things that are virtually guaranteed to be dead on arrival "across the way." The organizations that do respect the rules, do their homework, come prepared, and get down to basics may find the Congress receptive and often helpful.

On obtaining support for new technologies

One of the most effective techniques of promoting the use of new technologies has been the establishment by the government of small but stable marketplaces based on known needs in defense, intelligence, health, education, communications, transportation, and manufacturing. These markets are well received by a private sector glad to reduce (and help pay for) its risk in entering new product areas. The most impressive example of this technique in the past 50 years was in the development of computers and software in the 1960s. A remarkably small government investment "jump started" this development to such a degree that its market now far exceeds all government needs. This technique is well institutionalized in the Defense Advance Research Projects Agency (DARPA) with a remarkably wide scope, national defense, and all its needs. It has long had and merited the support of the Congress and the President.

Efforts by other agencies[2] to emulate DARPA's long and continuing success have been less successful, probably because their priorities were lower, their scope more limited, and the funding base of their department was less.

U.S. Government efforts to compensate for private sector decisions that allowed other countries to develop technology first by subsidizing U.S. industry (the famous high definition television (HDTV) and memory chip crises) were widely opposed and, as it turned out, ineffective. (American industry itself resolved the crises to its own advantage.) Even DARPA's efforts have been opposed whenever they appeared to favor or assist an existing industry; one such was hard rock mining research for underground shelters, opposed because whichever mining company was chosen for the research would gain a competitive advantage over its peers — so none was chosen.

The probable lesson learned was that government support of efforts that led to new markets that emerged from government needs, even if speculative, were successful. Assistance to existing private sector markets was not.

Excellent organizations looking for new fields and emerging or latent markets, if they can match those interests with those of DARPA, may well find "seed money" support. But not those who wish to compete in existing private sector markets. The belief is too strong, in both sectors, that the best private sector work in technology development, even "precompetitive," is done by the private sector based on its own interests.

The government as a sponsor of adventure and exploration

One of the most exciting and adventuresome functions of government is to acknowledge the hopes and dreams of its citizenry by sponsoring adventure, research, and exploration. Pericles, the leader of the Athenians in 453 B.C., commissioned the building of glorious temples on the heights above the capital city of Athens. The Acropolis, as it became known, was both a technical marvel and a celebration of the Athenians' recent victory over the Persians and ascendency into the ranks of national powers. The Greeks also had long been some of the earliest explorers and colonizers of the Mediterranean from the Bosphorus to the Straits of Magellan, as memorialized by Homer in the Iliad and the Odyssey, and established in history by major Greek cities all around its periphery.

Enfante d Henrique (Prince Henry) of Portugal sent ships around Africa so that Portugal would not be left behind in reaching the fabulous East.

Thomas Jefferson in the early 1800s bought and then called for the exploration of the Louisiana Territory, an astonishing achievement for such a young country. Countries from all over the world competed to explore the Arctic, Antarctic, the deepest oceans, and the highest mountains. The mountaineers themselves may have climbed the peaks because "they were there," but they also made sure all of us knew which country they represented — Everest for New Zealand, the North Pole for America, the South Pole for Norway (and Great Britain as an epic of courage), and the moon for the Russians (first instrumented landing).

In the most recent such demonstration, President Kennedy called for men to go to the moon and back in less than 10 years. In joyous pride, the astronauts announced on achieving the first part of that challenge, "The Eagle has Landed," as that eagle, the Lunar Lander, safely touched down on the lunar surface. Well, that eagle, too, may not be able to swim, but it did fly to the moon and return with its mission accomplished.

A good question about these dramatic explorations is not so much why they were done, but where they stood in the general priority of things. In particular, what were the reasons, in priority order, that justified their effort and expense? And how can those reasons help organizations interested in such ventures move in their direction?

History, from the projects of Pericles of Greece and Enfante d Henrique of Portugal to Presidents Jefferson and Kennedy, suggests a rough priority list could be compiled such as that the author has shown in Table 4.1, starting with the strongest national justification and greatest funding level (say, 10) and proceeding to the weakest one with the least funding (say, Level 1). Other headings and levels of importance for the second column might be the ratio of votes in the legislature, the relative interest levels of the population in the project, and so on. The reader should feel free to assign different numbers in that column based on other parameters. Regardless of their assigned relative values, the priorities on the left would stay about the same.

Table 4.1 Justifications Determine Support

Justification	Priority as Measured in Relative Funding Levels
Survival against threats or competition	10
Sovereignty and independence	5
Demonstrations of national spirit and strength (occasional only)	3
Science and the pursuit of knowledge	2
Peacetime sustenance of capabilities essential for war	1
Total	**21**

Note: To determine the possible relative funding level for a specific program, check which justifications are applicable and sum their relative funding levels. Clearly, the more justifications, the greater the national need and the greater the likely funding. Examples from the text: Cold War = 18, Apollo lunar landings = 8, Basic research = 2.

The Cold War, for example, resulted in Level-18 funding for developments in fields of military value (10 + 5 + 2 + 1). The driving reasons for the American exploration of the moon were to "beat the Russians" and to deny claims to it by other countries (Level 5 + 3 = 8). Parenthetically, adding science as a justificatiion for the Apollo lunar program would not have made that much difference (+2) in whether the exploration was carried out — something to remember in considering the exploration of Mars. That science led to a better future justified Level 2 research grants to individuals and institutions based almost solely on belief, not on its potential payoff. For decades between the World Wars, Navy oceanography sustained facilities and personnel in peacetime essential for Naval prowess in war (Level 6+). When oceanography was transferred to the National Science Foundation, it dropped to Level 2.

The value of such a priority list to excellent companies looking for new fields is in estimating ahead of time the value judgments to be made by the Congress in funding them, depending on the totality of its justifications. In the early 1960s going to the moon was supported by the second and third of these justifications. But by the late 1990s, 30 years later, interests had changed and it became difficult to finance the costs of the space station or manned flight to Mars even assuming the full value (6) of the last three. Automated scientific exploration of the solar system now rates only the fourth justification or a funding count of 2 or so.

Therefore, in looking for adventurous new fields, tally up the funding levels of the possible justifications before going too far. It is a primitive test, of course, but easily applied. One of its first uses was in about 1960 when a politically astute NASA official visited the NASA/Caltech Jet Propulsion Laboratory to help explain to its "spaceniks" why the funding of our dreams of planetary exploration was going to be limited (and was). By the time he had sketched the equivalent needs and stakeholders on a chart, it didn't take more than a glance to see we weren't very high on anyone's priority list. We never bothered to fill in the numbers. A good lesson, well taught, by an understanding sponsor.

The government as a trading partner

One of the primary tasks given the Executive Branch of the Federal Government by the Constitution is foreign affairs; that is, the management of relationships with other countries. The management of economic affairs and international trade, previously a relatively minor part of the task, has now become a vital one for two reasons. The first is the globalization of the marketplace with its biggest single market, that of the U.S., now open to world competition. The second reason is because of the unprecedented efficiency of international trade brought about by three agents of change — global satellite communications, global air transportation, and global information networking.

The three agents of change

One of the more beneficial results of the Cold War was the accelerated development of communications and surveillance satellite services. Because of their developments, people all over the world can now observe *and react to* events as occur almost anywhere from the bottom of the sea to the edge of the solar system, from riots in the streets to boardrooms and stock exchanges, from individual victims to heads of state. Unrest in a remote capital of a small country can result in decisions to pull out of major investments and organizations in that country within a few days, sometimes within a few electronic seconds. Billions of dollars can be moved anywhere at the touch of a button.

A second development was that of relatively inexpensive air transportation, making it cost effective and profitable to move high-value goods, from foodstuffs to computers, from maker to seller to buyer within a few days.

The third development was in trusted, secure information processing. Orders, confirmations, and payments can now move reliably and safely at the speed of light over satellite, microwave, and cable links. Computerized access to an extraordinary amount of information and knowledge has made problemsolving easier, more accurate, simpler and faster by orders of magnitude. Credit ratings, criminal records, background checks, stock values, technical specifications, even satellite intelligence, can be accessed in seconds for millions of industrial, business, government, and personal users.

Less visible are the societal and governmental changes that have come about in response to these developments. The same technologies that led to global communications and transportation are now making global organizations not only practical but imperative. Further, not only can manufacturing, financing, and other corporate functions be done anywere, so can headquarters be located anywhere and nowhere — in the sense of being instantly relocated by electronic transfers to redundant databases. Indeed, this capability is becomng mandatory for these organizations to continue to function during and after natural disasters and wars. Mergers are occurring based almost solely on the belief that unless an organization is active in all three major markets — American, European, and Asian — they cannot succeed. Such organizations can control more assets and people than most of the nations of the world and yet avoid being under the total control and regulations of any one of them. Many companies, such as the commercial shipping industry, have for decades chosen to be registered, chartered, and headquartered in whichever country or state they choose and gives them the best deal. Economically, the incentives to form larger and larger organizations and cartels are so strong, and arguably so good for much of the world, that they will prosper and, in any case, not easily be dismantled. Their constituency is just too large for that.

The effects of these changes on national governments as trading partners

An old saw in diplomatic circles is that the essence of diplomacy is delay. Pressing problems may go away. So, give people time to think them over before prematurely coming to a decision. During the centuries when even the fastest delivery of official documents could take weeks, there was a ready rationale for deliberate delay. The original Constitution, itself, allowed 4 months to elapse from the time of election to the time of inauguration of a new President based on the time it took to go from place to place by horse carriage.

Today, the American electorate knows within hours of the closing of voting booths which individuals have been elected President, Senator, judge, or village mayor and what to expect from each of them. Technologies have now made such delays technically unnecessary, even dangerous, as lame duck representatives pursue interests explictly rejected by the electorate.

Shuttle diplomacy, an effort to accelerate progress between a few recalcitrant nations, has become the norm, to the near-exhaustion of the shuttle diplomats. Personal shuttle diplomacy involving a dozen countries, as was the case in the Kuwait–Persian Gulf War, is fast becoming impractical. Shuttling among inter and transnational companies that can move assets electronically around the whole globe in seconds can be an exercise in frustration.

Changes by one government in its antitrust, environmental, public utility, human rights, and financial laws can be offset within seconds by another well-alerted government. The policies that once favored a nation now became toothless, their originators baffled. Defense policies aimed at national self-sufficiency of critical materials and technologies lose their effectiveness as more cost-effective parts from other countries find their way into critical communications, transportation, manufacturing, and military equipments. Trading is no longer as simple as it once was when individual countries controlled their own markets and industries, when national boundaries were absolute, and when powerful countries could enforce their own will. For better or worse, the U.S. Government is now less able to either assist, regulate, or sanction its own markets, industries, and economies — much less those of other nations.

The reaction of nations to global economic change

The nations themselves, in order to decrease the resulting instability in world economics, have begun to work more and more with each other in areas of common interest — crime fighting, intelligence gathering against terrorist groups, environmental protection, and transborder trade. Two of the many alliances that have been formed recently are the NAFTA (North American Free Trade Alliance) and the European Community, both dedicated to resolving relatively uncontroversial regional problems. Success to date has encouraged others to join them. Trading has measureably increased among them and to some extent diminished the associated problems of illegal immigation and drug traffic. Tourism has become far simpler, enjoyable, and profitable.

For excellent companies with global interests, national alliances also can mean significant gains in efficiency and lessening of uncertainties and internal interfaces. They do, however, require real understanding of the new business cultures that are emerging in these alliances, separate from the cultures of the individual nations. It will take a bit of architecting, but alliances should open up new, less risk opportunities for companies prepared to make the internal changes that will be required.

Summary

Four roles of government directly affect the success of excellent organizations contemplating change — national security, regulation, purchasing, and foreign affairs. All require governments to maintain a delicate balance of competing interests at the cost of delay and uncertainty of commitment. All became more complex and difficult to manage in the last decade of the twentieth century.

The essence of a stable government has long been well-crafted, strictly followed rules — a discipline hard to maintain during periods of rapid change.

The U.S. Government, particularly the Department of Defense, is restrained by the Constitution and the Congress, both by law and through civilian control and funding. Regulation is the most contentious of the roles which tends to swing back and forth between over-regulation and excess deregulation, driven by the conflicting demands of the public and private sectors. The Government is a sizable customer of the private sector, but it operates to its own set of rules, knowing and respecting them is essential. Adventure and exploration remains one of the most unifying roles of government shy of a "good" war.

Perhaps the roles most affected by the revolutions in global communications, global transportation, and global information processing are the managements of foreign affairs and security in both the public and private sectors. The end to the Cold War and the rise of instant reporting has drastically changed the face of war. Deliberate delay, the standby of diplomacy, is now impractical. National control of international organizations is rapidly diminishing.

In relatively stable if dangerous periods, such as the last half century in the well-developed Western countries, excellent companies have proceeded with what they believed was a good course of action without worrying too much about what governments might do, intentionally or not, that would preclude it. Rules, for better or worse, were stable and predictable and there

usually was time to change new rules if and when unintentioned consequences began to occur. The companies were reasonably sure of the nature
of foreign markets and could count on governments to assist them in entering
and working in them.

These assumptions are becoming less and less true as the global economy
increases in importance and with it the influence of nations decreases. It
remains to be seen whether the freedoms gained by excellent organizations
in all countries will be worth the loss of military and economic protection
that national governments once could provide them. In the meantime, the
formation of national alliances for economic purposes seems to hold real
promise.

When all is said and done, the U.S. Government itself is one of the most
knowledgeable and effective partners of excellent organizations in times of
radical change. National security is acknowledged to be the best managed
of all governmental roles. The excesses of over-regulation are being reduced.
Appreciably reduced in purchasing capacity, the Government is less of a
factor in private sector intervention than in the Cold War era. Indeed, from
the standpoint of making radical change, it is less risk to move from the
private sector to the public than the reverse. The latter, for better or worse,
performs by the rules.

Notes

1. *Websters II New Riverside University Dictionary.* The Riverside Publishing Company, Boston, MA, 1984, 990. Unless otherwise stated, all word spelling and
   useage is from this reference.
2. The National Science Foundation, NASA, the Department of Transportation,
   the Department of Congress, and even other countries.

Part III

*Internal constraints
and opportunities*

Six eagles in a row that couldn't catch their fish

Nothing fails like success. (Frank Capra, ca 1980)

Introduction

In the mid-1980s the author participated as a witness to a strange story, one that had a profound lesson to teach, a lesson which a dozen years later became Part III of this book. The story has a metaphor: six eagles that in almost clockwork sequence failed to catch the fish they expected to have for dinner. The eagles, in reality, were some of the finest aerospace companies in the world. And the fish were major contracts they expected to easily win. The lesson was that even in excellent companies, sound objectives, policies, structures, and management can prove to be weaknesses when it comes time to fish in a new setting.

The story of six eagles in a row that couldn't catch their fish

This story begins in the early 1960s when a young agency in the Office of the Secretary of the Department of Defense, the Advanced Research Projects Agency (ARPA), was tasked to develop a series of Earth-orbiting satellites for the military purposes of communications, navigation, weather observation, remote sensing, space exploration, and what later became intelligence and strategic warning. Their design and construction was duly contracted to a series of aerospace companies including The Boeing Company, General Electric, Hughes Aircraft, McDonnell Douglas, Rockwell International, and TRW — eagles of companies, all. Within a relatively few years, all had created some of the most remarkable space machines ever imagined. All of them more than accomplished their assigned missions. Indeed, using the criterion of "bits per buck" (information generated per dollar), all companies within a decade or so exceeded all expectations. U.S. Presidents acclaimed the satellites, pointing out that they had paid for themselves many times over just in the reduction of national security costs. Possessed of clear and timely information, leaders could better plan programs and resources, avoiding the need for a worst-case design. Fortunately, the Soviet Union was able to, and did, take a similar approach with similar results. In the American case, because the satellites had been so well designed and built, their life on orbit approached an average of 10 years or more, two and three times longer than planned — and consequently less costly per mission.

It was a quarter of a century later before the U.S. Government felt it necessary to replace these magnificently architected systems with ones that, given new technologies, were anticipated to be still better in cost effectiveness, weight, size, and survivability to enemy attack. Requests for proposals for each satellite type in turn were written, commented upon, and distributed to all the companies now very successfully engaged in the satellite system business. The bidding process was scrupulously honest and fair, the playing field as level as decades of experience could make it. To illustrate, questions by a bidder during the bidding process were answered quickly by the government, with both questions and answers distributed to all bidders.

Now the story takes a surprising twist. When the proposals for the first of the new satellite systems were received, evaluated, and the winner announced, the company that had built the first generation of this type *lost* the contract to another bidder. Well, that happens.

Essentially, the selection process was then used for the next, quite different, satellite system. Again the loser was the original builder. Although there were some raised eyebrows, there were no readily apparent connections between the two surprises, and no questions or comments raised by the competitors.

After six original builders in a row had lost the bidding for "their" systems, it was obvious enough that something fundamental was occurring. All the usual suspects — politics, conflicts of interest, inexperienced bidders, and a mismanaged process — by now could be ruled out.

Parenthetically, on-orbit results subsequently confirmed the soundness of the judgments of the evaluators and selectors. The selected architectures clearly were better on a number of counts than those of the first generation systems they replaced.

It was evident from the bids, however, that all the original builders had used an evolution of their first design as the basis for the proposed one. There were good reasons for this, to be sure. Each original builder had all the resources, background, experienced people, jigs and fixtures, design tools, and specialized government contacts to do the job well. Each certainly knew more about the capabilities of their specialized satellites than any other bidder possibly could. Each certainly planned to win. None did.

The trail to find out *why* none won quickly led to the beginning of the bid and proposal process. The usual first step is the issuance of a draft request for proposal (RFP) about a year before the final RFP is issued. In any case, all bidders were given the same 90 days to respond to a final RFP, with late replies to be ruled unacceptable, without appeal. Now, it is a fact of life in proposal writing that, given 90 days in which to respond, the first 30 are spent in refining and cross-checking the proposed design and the remainder in satisfying the voluminous procurement provisions. In this case, all bidders satisfied the latter, so whatever happened that differentiated the proposals one from another had to happen, or be committed, within the first 30 days. Verily, as in many things, including proposal writing:

> *The beginning is the most important part of the work.*
> [Plato, fourth century B.C.]
>
> reformulated recently as
>
> *All the serious mistakes are made in the first day.*
> [SP 88 and RE 91 48]

This much of the story is fact. The resolution, reached in Chapter 5, was found to apply well beyond this particular story and is extended in Chapters 6 and 7.

chapter five

Unstated assumptions and other subliminals*

> *The most dangerous assumptions are the unstated ones.*
> [KI 91a 13]
> *All the really serious mistakes are made in the first day.* [SP 88]

Introduction

The Introduction to Part III began the story of how six different "eagle" organizations in a row lost bids to build a second generation satellite system to one it had built and operated highly successfully for years. This chapter first concludes the story and offers a likely reason for its unexpected conclusion The chapter then focuses on unstated assumptions and other hidden barriers that need to be surfaced and challenged before an excellent organization can effectively change its direction and remain excellent.

The six eagles: The search for causes

That the original builders had decided early in the bid process on an evolutionary design was evident enough from the designs they submitted. But it also was easily predictable by the other bidders. For them it made no sense to propose something similar. If they had, they would have lost against a more credible bidder, the original builder. They had to re-think, to re-architect, to start "with a clean sheet of paper." That they did and won.

A first quick look thus would seem to show that the cause of the lost bids was the choice of an evolutionary design instead of a revolutionary one and that a complete redesign was a better choice. But,

> *The first quick looks are often wrong.* [RE 91 152]

---

* Subliminals: factors below the level of conscious perception. [WE 84] and this was no exception.[1] Was the evolutionary approach a possible contributor? Perhaps. Most of the time evolution is the preferred, conservative, predictable approach. After all, the purpose of complex systems is to create a new capability, not necessarily to use the highest technology available.

Conversely, was re-architecting necessarily the right choice? Not necessarily, even if a quarter century did occur between the generations. In point of fact, analyses by The Aerospace Corporation show that the greatest single reason for cost and schedule overruns is the *premature* use of the very technologies that make many new architectures possible. Reason enough to be cautious. But the clincher was that at least one of the winning designs was not particularly revolutionary or even high tech. It won by an insight.

Winning by insight

Although each of the winning archiectures was for a quite different purpose, one of the winners was notable, not for its technology, but for the insight it showed into the *real* problem.

In this particular case, the Congress as its client required the new satellite system to meet what it perceived to be almost identical needs of four incumbent systems built by four different space-oriented agencies. The idea of a single satellite system to meet everyone's needs was accepted, but with great reluctance, by the four agencies which were proud — and protective — of their own highly successful, first generation systems. The winner acted on this political fact, obviously not mentioned in the RFP, and designed its satellite architecture based on four "wings" attached to a central core. The core provided the essentials: power, housekeeping telemetry, command channels from the Earth, and so on. Each wing met the specialized needs of an agency, one wing per agency. It was a master stroke and far more than a political gesture. It assured sufficient independence of each agency's programs that the proposed performance, cost, and schedule targets were credible and likely to be met. Mutual interference had crippled more than one program before. As might be expected, that architecture won the bid with the enthusiastic approval of the agencies. More importantly, it met its promised targets. The fundamental principle?

An insight is worth a thousand analyses! [SO 93]

A telling sequel

There was an epilogue to this story that is at least as instructive. Within months of the contract award, Congress stopped all funding of one agency's program. By simply substituting a properly designed dummy wing, the rest of the agency programs continued largely unaffected. Some months later, another agency appeared on the scene, desiring a "ride." It would procure its own wing but otherwise not interfere with the other three passengers. As a matter of fact, none of them would even see their fellow passenger before all were assembled on the launch vehicle. Further, this passenger would be turned "off" for all ground tests and would turn "on" only when the satellite reached orbit. Ten years later the other participants were quietly informed that "the fourth wing had worked as intended and thanks for the ride."

Had the original, tightly integrated architecture been chosen, none of these post-contract changes would have been possible and the whole program might have collapsed.

What other principles were demonstrated? First, a general one:

> *Don't assume that the original statement of the problem*
> *is necessarily the best, or even the right one.*
> [RE 91 54] [R&M 97 26]

and second, a more specific design guideline, certainly in this case:

> *In partitioning, choose the elements so that they are as independent*
> *as possible; that is, with the minimal communication*
> *between the subsystems.* [RE 91 41] [R&M 97 26]

This second heuristic is seen in many guises. Very popular, it has at least five variations. [RE 91 312] [RE 97 150] In technical programs, it helps let subsystems be developed more or less independently, not getting in each other's way. In organizations, it calls for strong decentralization, focused responsibility, and clear accountability. It certainly was well used in our story.

Unfortunately, winning by insight may be wonderful, but it is hard to plan ahead. In this case it helped show that winning wasn't always a matter of a decision to re-architect. But if evolution weren't the cause itself and, yet, was present in all the losing bids, then it must be a consequence of something still deeper, something not so easily fixed, a *condition* leading to evolution rather than a single decision.

Can't win for losing

As one RFP after another went out and one after another of the original builders lost, this strange result must have become more and more apparent, especially to those who had built the systems yet to be procured. And yet the pattern still didn't change, even after six in a row! It seemed that the remaining original builders couldn't win for losing. Clearly, they must have felt very strongly that an evolutionary approach was the *right* one or they would have abandoned it for no other reason than, for reasons yet to be understood, it was a losing proposition. But they didn't abandon it and lost — again.

Indeed, they still might have won, evolution and all. One strategy applicable to their case is suggested by a heuristic in Chapter 3; namely,

> *For every competitive system there will be*
> *at least one countersystem.*

That is, just as the winners of the contracts predicted the original builder's evolutionary strategy and architected to beat it, the incumbent builders might have predicted their competitors' re-architecting strategy and taken steps to beat *it*. For example, the incumbents might have set up a design team (a so-called "red team") to try to beat their own new and improved evolutionary design. Or they could have offered two proposals: one evolutionary, one re-architected. Or they could have modified their evolutionary approach to one offering more options to defeat at least some of their competitors. For example, if their original software architecture were a closed architecture (a likely possibility in the 1960s), they might have changed it to an open one, greatly expanding the options available to a user. They still might have lost, even with a red team, but that team might have been able to tell them *why* they lost; in short it was because:

> *The incumbent carries the encumbrances.* [WI 91]

All architects, designers, and politicians know that no system or person can satisfy all the needs all the time. With strengths in one area almost always come weaknesses in another. The deficiencies of the incumbent system or politician, however, are on display from past performance and most stakeholders and challengers focus on getting rid of them. The incumbents, before even beginning to criticize their competition, must mitigate, minimize, or eliminate their own encumbrances. Even better, they can show new capabilities and options that only the original design can produce. But first, acknowledge and handle the encumbrances.

Whether these strategies were or were not used is uncertain. What is certain was that the incumbents were unable to abandon their original architecture. An increasingly accepted guideline suggests why:

> *The exceptional team that created and built a presently successful*
> *product is often the best one for its evolution — but seldom for*
> *creating its replacement.* [R&M 97 242]

In other words, if the proposal team consists of or is managed by the people who created and operated the earlier and sucessful design, their next design is most unlikely to commit to one which abandons the basic ideas of the earlier one, and especially in only 30 days. After all, to do otherwise could be perceived as rejecting as flawed what some of the team members had spent a career in building. In our story, the original builders did not abandon the evolutionary design because they couldn't. They might even have had a justified belief that the RFP could be undercut, as many have been, by a client decision that the new system would be too expensive or risky and that it would be better to stick with the incumbent one for a few years more. If so, then the incumbent builder couldn't lose.

For whatever reasons, the original builders were locked into ideas and assumptions which were so ingrained that they didn't even surface to be recognized, much less challenged. "Of course that won't work. We tried it several times over the past 25 years and it never did."

The author, in rare moments, admits to the same self-conditioning. As the initial architect of the NASA/JPL Deep Space Network (the DSN), a radio system for communicating and navigating anywhere in the solar system, I "know" that there still is only "one best way" to build such a system. In actuality, the DSN has made so many significant additions to its architecture that the original one, while still there as a highly reliable system on its own, could not have produced some of the more dramatic "firsts" in comparative planetology.

On abandoning an obsolete architecture for a new one

A truly remarkable abandonment of an architecture, a circular waveguide system for long distance communications, occurred at the Bell Telephone Laboratories (BTL) at about the same time. It was remarkable because the abandoned architecture had yet to see service! Stewart Miller and his research team abandoned it after 15 years of research and development effort and just before it was to be demonstrated in an operational link. They opted to start over and develop a fiber optics system, *years* before the enabling technologies for it existed. There were no technological solutions to the connector, optical fiber materials, and wideband switching problems even in sight! To Miller, the end game had to be fiber optics and the team aimed for it before the prior technology (circular waveguides) had even been installed, much less evolved and improved. As history later showed, the decision was not only courageous, it was right. It was a major factor in bringing optical fibers into use not only in telecommunications but in fields from orthopedic surgery to woodcraft. And the team became heroes.[2]

Abandonment can be a very emotional issue; the abandoned system is one's child. It can be a troublesome management issue. Changing a product architecture almost always means changing the organization that has supported it, including the people that did so with the most intensity and success. It is even harder if the top management reached its height based on having created, developed, nurtured, and fought successfully for it.

Yet there may be no choice. Your research group, your competitors, your long-range planners, your new CEO, or your customers may already have a new one in mind. Never mind that "it won't work." You still have to give it your best shot. Who knows, you may hit (on) something and save the day.

Be that as it may, it is too soon in this book to tackle this subject head on without more preparation by Parts III and IV. This section asks only that the reader take heed. There are more minefields ahead.

Assumptions of the nature and scope of success

The nature of success and failure

Excellence and success, as was noted in Part I, are not the same. The story of the six eagles, all excellent but unsuccessful in this instance, is a case in point. Indeed, one of the most damaging mistakes that organizations and individuals make is to equate the two, to assume that excellence means success. The consequences can be far greater and longer lasting than losing a bid.

Two of the finest laboratories in the U.S., the Naval Research Laboratory (NRL) and the NASA/Caltech Jet Propulsion Laboratory (JPL) were almost destroyed, their futures compromised for years, when NRL's Vanguard satellites and JPL's Ranger lunar probes [HAL 77] failed catastrophically in front of a world audience, not just once but time after time.

Prior to these disasters, NRL had a decades-long record of accomplishments in naval electronics and radar; JPL had a similar history with the U.S. Army. In addition, it had a major part in successfully launching America's first satellites using the Army's Redstone rockets, discovering the Van Allen belts in the process.

Then disaster (investigations, recriminations, punishment, and destruction of careers) were again on the public stage. The honor of the U.S. had been sullied and the Congress was enraged. JPL managed to recover its luster with later flights to the moon and to all the planets but one. But it took decades. NRL never again reached premier status in space. NRL's subsequent achievements, particularly in space science, surveillance, and naval support were seldom given the public recognition they deserved.

Robert J. Parks, then a young supervisor at JPL, expressed the lesson learned this way, "We learned that putting a JPL sticker on the side of something doesn't mean it will work." The lesson transformed that laboratory (in the long run, very much for the better) into one of the most disciplined and professional systems organizations in the world. Whatever hubris and self-congratulations might have existed before had vanished.

But as a more bitter statement at the time went, "When you succeed, you shake the hand of the President. When you fail, you're a bum." Such, indeed, was the fate of some of JPL's finest engineers and managers. Regardless of the cause — to be determined perhaps years later — failure is unacceptable *immediately,* and the greater the excellence of the organization, the greater the fall from grace. Such is the nature of success and failure. Ask NASA and Thiokol, the manufacturer of its solid rockets, about the *Challenger.* Ask Dow Corning about implants. Ask Ford Motor Co. about the Edsel — a symbol of failure — named, with great hope, after one of the finest engineering members of the Ford family. Ask Lockkheed Aircraft about the Electra, the hoped-for successor to the world-famous Lockheed Constellation, the "Connie."

Yet, all of these organizations were top-of-the-line, excellent in anyone's book.

The scope of success or how did we get in this fix?

Not the least of the reasons for disastrous failures is expectations which exceed what is possible. As one saying goes:

> *High expectations, because they are unlikely*
> *to be fulfilled, define failure.*
> or, in the vernacular,
> *Their reach exceeded their grasp.*

Which might make sense in 20/20 hindsight, but would be unfair to those involved. Certainly in the cases just mentioned, no one deliberately planned unreachable goals — though such goals can serve useful purposes elsewhere.[3] Nor did they opt for the opposite, self-serving approach:

> *Low expectations, because they are likely*
> *to be accomplished, define success.*
> or, similarly,
> *Never let your reach exceed your grasp.*

In the extreme version of this approach, one never announces one's objectives, develops possibilities in secret, and announces the results only after they are "successful." In a free society of excellent organizations, this strategy is both impractical and self-defeating. The successes, even if they are astonishing, are tarnished by legitimate skepticism once the strategy is surfaced. This, too, is unfair to those who achieved the success. The manned space program of the Soviet Union is one of the most poignant examples. For shear courage and success against great odds, the Soviet manned space-flight effort can be compared only with the U.S. manned flight to the moon. Yet for decades the long-sustained Soviet manned space station program was virtually unknown or unreported in the West. It wasn't matched for still another decade. [OB 81] [HAR 97]

For better or worse, to be perceived as real successes, programs have to announce their objectives well before accomplishment and, if not achieved, they have "failed." There is no simple and easy answer to this often tragic conundrum.

How, then, do excellent organizations get into this fix? There are several causes. Unless expectations are sufficiently high, they won't attract interest and support, especially from governments. As work progresses, expectation-creep begins; more and more hopes and dreams spawned by progress to date keep raising the bar for success. In an oft-repeated metaphor, the young pianist plays a few Bach exercises well and the family dreams of Mozart concertos. When that begins to be possible, the dreams are of the Tschaikovsky Prize and world fame. But such fame occurs perhaps once every few decades. The still-young pianist fails, by definition. Or a budding professional sets personal goals so high that the result is psychological breakdown over not achieving them. Another cause is unavoidable dependence on others to succeed — say, a launch vehicle developer for a satellite program, a steel maker for a high-rise office builder, or a microchip maker for a new computer, all of which have occurred — and the programs dependent on them fail, publicly and catastrophically. Fortunately, as was said, this conundrum is not a new one. Architects have struggled over it for thousands of years, and they have found that it can at least be cut down to size.

First, they learned to appreciate that success is in the eyes of the beholders and there are many of them. In other words, regardless of whatever success is achieved there will always be those who perceive it as a failure, and vice versa. For architects, success is a satisfied client. A satisfied client is certainly necessary and that should be sufficient for the architect.

Second, they learned not to define success in any effort as being greater than what you (and your client) can afford to lose, be it financial in nature or professional reputation. To assist in that objective, there should be a feasible "Plan B" for recovery, if at all possible. In a tragic example, NASA created an impossible expectation, followed by an international trauma and severe damage to its hard-won reputation for success by choosing a school teacher as a Challenger passenger. Had all passengers been test-pilot astronauts...

And third, architects know the importance of defining success and how it will be determined *with* the client *before* starting and to *stick to it* with the vehemence of a good configuration control manager. In practice, this has meant that client satisfaction and acceptance need to be based on measureable criteria established very early in the program [RE 91 144] and not on unstated client hopes and dreams developed later. Passing acceptance tests then can give to the client, the architect, and the builder a limited, negotiable, and measurable definition of success with major impact on system design:

For a system to meet its acceptance criteria
to the satisfaction of all parties, it must be architected, designed,
and built to do so — no more and no less.
Furthermore,
Qualification and acceptance tests must be both
definitive and *passable.* [RE 91 241]

This last heuristic may seem strange on first hearing. Yes, it helps make sure the tests are realistic; that is, both measurable and achievable. But it does much more. The author was confronted with just how much more in 1961 when attempting to award a contract for a very large, unprecedented DSN antenna based on specifications that called for proper operation in a 100-knot wind during earthquake shakings of $1/4$ g in any direction.

The obvious after-the-RFP questions should have been expected. What does "proper" mean? That question was answerable. The antenna had to point in the direction of a spacecraft to within a specified fraction of a degree. How do we test this antenna in a 100-knot wind when such wind velocities occur in this location perhaps once a decade or so? And, "Let us know when this here earthquake can be scheduled for our test." The so-called performance test was physically impossible to carry out. Needless to say, the specifications were rewritten in the testable terms of "measurable structural stiffness." Sounds complicated; it wasn't.

For the nonengineers among the readership, "measurable structural stiffness" is determined by measuring the antenna's lowest resonant frequency of vibration, something almost literally done by banging on the antenna's movable frame with a sledge hammer and listening to the sound it makes. If the lowest measurable frequency was higher than one cycle per sec, the antenna could pass the 100-knot wind and 1/4 g quake requirements! It did pass and subsequently delivered as hoped and intended. Was the change in the requirement important? Had it not been done, the builder's only recourse was to insure the antenna from the beginning, doubling the pre-RFP estimate of about $10 million and terminating the project. On a larger scale, it would have reduced the data return from all future spacecraft by a factor of 10 and could have converted the DSN from being the product of a front-running excellent organization to a relatively routine engineering task of a support group.

In the long run, of course,

> *The test of a good architecture is that it will last.*
> (Spinrad, Robert 93)

This heuristic gives a perspective to success that is often forgotten — the difference between immediate test success and long-term utility. Acceptance testing assumes that the intended use of the system can be specified early in its development with some exactitude. Long-term utility recognizes that actual usage may be so different that actual success or failure may take some time to appreciate. It is certainly doubtful that Arthur Raymond, when architecting the DC-3 passenger airplane, fully appreciated its astounding success over a 50-year period in many applications under widely different conditions. It is unlikely that the global success of the GPS in civilian navigation and positioning use would have been predicted by its military sponsors and builders a decade earlier in the midst of the the Cold War. At the same time, architects and clients would not have predicted the failures of some sure-thing projects that succumbed to changing circumstances, either.

The answer to the conundrum? Success is satisfying the client by passage of acceptance tests which requires an early mutual agreement on what that will take to achieve it and how it will be measured.

So many unknowns and so little time

Almost by definition, any radically new venture begins with too little information to make well-informed, even sensible, decisions. Cost estimates are demanded which vastly exceed the data available. Schedules are projected without even knowing whether essential suppliers will be available or even exist. Materials and techniques are postulated which are still in the research phase. Unfortunately, in the beginning there is not much with which to predict their future in any detail. It is the time of the subliminals — of below-the-consciousness drivers, of ulterior motives, unstateable assumptions, and unsupportable assertions. A time to be wary, yet a necessity to be bold. The time when the really serious mistakes are made.

Some terrible traps

Those who have been in the first meetings of a long project-to-be have heard at least one of the following foolish arguments proposed in support of a new program:

> "Look at this huge market! All we need is to get 1% of it
> and we'll be rich."
> "If it can be imagined, it can be done."
> "If we can go to the moon, why can't we..."
> "If they can, we can..."

The first variation finds a well-populated home in the computer, communications, and entertainment markets. The second is the widely broadcast slogan of a major stock brokerage firm and, believe it or not, a Presidential policy in the 1980s which led, among other initiatives, to the mislabeled Star Wars program and its successors. The third helped initiate the war on cancer, evidently missing the unique confluence of circumstances and dedication that made the Apollo program an historic success and setting up expectations that would not be met for decades. The last one goes with every fatal underestimation of the competition, more a mark of the weak than the strong. Psychologists would see it as a symptom of insecurity, a put down to push up one's self.

The experienced listener has heard them all. They are traps for the unwary, the stuff of Dilbert satires. [AD 97]

Surfacing the inspirations

Then there are those lucky moments in "the first days" when the really serious *inspirations* are made — those sudden realizations which, based largely on fragmentary information, find the core of the problem and state it in clear, concise form, such as the "four agency" insight, the optical fiber vision at Bell Telephone Laboratories, phase noise instead of frequency jitter at JPL, the lean production system, and the two-engine DC-3 for passenger comfort. From that instant, all other arriving information seems to fit in place. Data inconsistencies now seem to make sense. In more generalized form and given enough examples which support them, the sudden realizations become not only guidelines to success in the project but insights for others in their own successes.

Notes

1. Whether searching for answers for failure in organizations, Challenger space-flights, aircraft crashes, or in other complex systems, this heuristic remains the same. Complex systems fail in complex ways. Demonstrating that there is only one cause and no other takes far longer than jumping to an expected failure as the answer. Generally speaking, if the failure were expected, it would either have been avoided or fixed.
2. This BTL story is courtesy of Max Weiss, a one-time member of the BTL staff.
3. Notably, to inspire the effort or to keep hopes high for long-term goals as long as intermediate success keeps coming along. More than one project has "lucked out" by the arrival of just the right technology, circumstance, competitor's mistake or capability at the last possible minute. As more than one beleaguered project manager said during the Cold War when the Soviets did something threatening just before a vote by an appropriation committee in the U.S. Congress,"Well, the Soviets saved us again!"

chapter six

Different structures, different rules, and, consequently, different capabilities

Introduction

Organizations can be viewed from many perspectives. Chapter 1 views them as complex systems, Chapter 2 as creators of emergent values, Chapter 3 as competitors, Chapter 4 as partners with government, Chapter 5 as sets of beliefs, and Chapter 7 as decision sets. This chapter views them as structures. Each perspective highlights a different set of opportunities and constraints that affect what excellent organizations can and cannot do, and why.

Organizational structures can be categorized in several ways, as well. One of the most useful and interesting is by their primary objectives: bureaucratic (to follow established rules), profit-seeking (to increase the bottom line), and cultural (to create a compatible team). Each type does some things very well and others with difficulty. All contain excellent organizations and dedicated professionals, a fact the author personally can affirm from working in and for all three. But the same rewarding experiences also have brought with them a feeling of sadness and dismay as each is observed criticizing, demonizing, and denegrating the others to no purpose. All are necessary, though, in different ways. All need each other. The reader, therefore, is requested to view them with this point of view in mind, at least in reading this chapter.

There are few if any "pure" bureaucracies, profit seekers, or cultures. Each overlaps the others. To some degree, each contains something of the others. A profitseeker needs an administrative bureaucracy. A successful bureaucracy is also a culture, one characterized by noneconomic motivations. Conversely, a successful culture needs a bureaucracy and a revenue generator (a "profit," legally speaking) to sustain itself. Nonetheless, the three are different and these differences determine what that organization, or part of it, can or cannot do well. That is, different possible directions for an organization will require different types or sets of types of organizational structures. The descriptions that follow begin with purpose, amplify with structure, and close with their roles in making a change in direction.

The chapter itself begins with bureaucracies, the world's most successful organizational form judging from size, numbers, services rendered, security for its employees, and responsiveness to orders. Virtually all organizations are in part bureaucracies because they are the structure of choice of any group performing a service either for others or internally for itself.

Using the description of bureaucracies as a baseline, profitseekers and cultures are then contrasted with it.

Bureaucracies: you succeed if you follow the rules

The purpose

Webster defines bureaucracy in several ways. The first one, and the one used here because it is structural, objective, and characteristic of excellent bureaucracies in particular, is

> "Administration characterized by *specialization* of functions
> under *fixed rules* and a *hierarchy* of authority."

The key words have been emphasized. Specialization implies expertise, fixed rules imply discipline, and hierarchy implies structure. The founding purpose was to create a division of labor into manageable parts in the interest of efficiency and control of the whole. Centuries of experience have shown that a bureaucratic structure is best used when satisfying a "noneconomic" need (defense, education, exploration, police protection, revenue collection, and business administration), that is, where value created is not easily quantified and where profit, as such, is not an objective.

The noneconomic needs that bureaucracies satisfy carry with them bureaucracy's greatest deficiency: the lack of a quantitative, measurable incentive to improve its efficiency and flexibility. There are many disincentives to becoming worse, of course — Congressional investigations, punishment by budget reduction, media attention, public outrage, and the like. Personal awards, medals, citations, and commendations recognize the merit of individuals and teams, and increases morale, but improving the organization's overall performance means proposing and changing the rules. That, internally, is almost heresy. Proposing and changing rules are actions reserved for the topmost echelon. The result is a structure which is inherently and *properly* conformist at its base and inventive at its top. It is at the top where rules are made and then promulgated as "guidelines." No one in the middle or lower echelons, however, would ever treat such a top-down "guideline" as anything but a command; one seldom deliberately violated.

The structure

A bureaucracy partitions the services to be performed into units or "bureaus," each containing specialized professionals strictly following the rigid rules of their bureaus and professions. That strictness is both a strength and a weakness of bureaucracies. It keeps order but resists change. Once energized, the momentum of this structural form in getting things done is awesome; once dismantled — as was the infrastructure of the Soviet Union — its specialized parts can no longer survive alone. It provides service but only according to the rules. Some of the strictest are

- Obey the law at all times.
- Follow the rules, don't question them.
- Follow Robert's Rules of Order in controlling contentious debate.
- Particularly, obey rules against discrimination, nepotism, conflict of interest, breaking the public trust, and endangering public safety.
- Implement Presidential Directives responsively and expeditiously, e.g., President Johnson's affirmative action program and President Truman's desegration of the armed forces.

Establishing and maintaining strict rules, regulations, procedures, and laws does demand an impressive amount of documentation. But paperwork is the lifeblood of a rule-based organization. Paperwork, too, has rules. For example:

The mail must be read and acknowledged on arrival.
(A.k.a., the "Roger" rule for telephoned orders.)

For every critical memo there must be a response memo.
Lack of same implies consent.
(A.k.a., one of several CYB (cover your backside) rules.)

All rules and specifications should have "bail out" or
waiver provisions.
(A.k.a., it's foolish to box yourself in. No rule
can cover all circumstances.)

An order is not an order until it is in writing
and signed by appropriate authority.
(A.k.a., Be brave. Be the fall guy.)

The author of a document expects all those
on the distribution list to read it.
(The other side of the first rule.)

This last rule also is an imperative in systems architecting and engineering. Systems are so complex and interactive that new information must be distributed to affected elements quickly so that appropriate action, including objections, can be taken to keep the system from being damaged. Failure to read a document from someone else is not only a disservice to the sender, the recipient may lose an opportunity to take action in time. There is no place in a bureaucracy for "handshake" agreements.

The last rule also can be the basis for a quiet, tactful reprimand called the "bank shot." A bank shot is a memo sent in one direction (to a named recipient), but is really intended for another (to someone else, notified by being on the distribution list). For example, a memo critical of an undesired action by just one subordinate can be addressed to "All Concerned," including the miscreant. Thus, everyone is alerted but only one is being called upon to shape up *now.*

Make no mistake, strictness in following rules is an absolute necessity. It is exactly what the people it serves want and must have. Creative accounting, idiosyncratic legal judgments, disobedience of military orders, preferential personnel administration, and informal security controls are only a few of the ills that bureaucracies and their strict rules are built to suppress. Therefore, for many good reasons, a bureaucracy is the organizational structure of choice for the military services, governmental agencies, service organizations, and the administrative side of profit-seeking companies.

Generally speaking, bureaucracies are structured as hierarchies (see Chapter 1, Figure 1.3). Staff is always shown to the side, one's boss is always "above," and dotted lines are rare. Solid lines mean reporting in one direction and control in the other, and nowhere else. Control is a critical factor in bureaucracies, not only in making sure subordinate elements are working properly, but in makng sure that their immense assets and power cannot be misused by those in charge for other than their intended purposes. As was shown earlier, Congress needs to be sure that the Defense Department (the military-industrial complex) cannot take over the government. The tools are laws, regulations, appointments, audits, inspections, and reports, which explain the intense interest the Congress displays in them.

There are other tricks of the trade, too. On how to make a respectful response to a dumb idea, "I agree, but there are some details to be taken care of before implementation..." On announcing bad news: the "sandwich" letter in which good news is presented in the beginning and the end, with the bad news in the middle. Or the expected response to or by a Congressman to an undesired question: answer, but interminably on a different subject.

As to be expected in bureaucracies, their members are judged primarily on their adherence to the rules, followed closely by how well they use them. Personal success is and should be in following the rules, not changing them. Following the rules might make being a professional in a bureaucracy dull and uninteresting except for one factor often forgotten by the outside world: providing a needed service is a *calling*, not a job. Defending ones country, helping people, educating students, and so on are causes which inspire dedication, a sense of personal value, and pride — even under conditions that, absent cause, would lead to mass resignation. This reservoir of dedication is one of the great strengths of bureaucracies, one not to be taken too lightly. There is more to being in a bureaucracy than simple job security. There is a sense of belonging, that the organization, particularly the military, will take care of its own, that you are among others that deeply share the same beliefs and causes. In that sense, a bureaucracy is very much a culture. As a culture, it can be moved by an appeal to its beliefs, or be an adamant opponent to whatever challenges those beliefs, and with remarkable alacrity. For those who would change excellent institutions, this can be a powerful drive, in either forward or reverse direction.

The inherent stability of providing noneconomic services — as opposed to the inherent dynamism and transient nature of profit-seekers — means creating a valuable long-term outlook in these days of short-term success and failure. Bureaucracies, by investing in facilities over long periods, can in their way be remarkably cost-effective. For example, only a bureaucracy could have built the dams, waterways, highways, and infrastructure that opened the West in the U.S. and much of the rest of the developed world.

Only a bureaucracy could have, under a single Presidential order, desegrated itself. Only a bureaucracy could go to war, on command, without complaint nor question. A story, probably apocryphal, is told of the then-Admiral of the Sixth Fleet aboard a ship in the Mediterranean Sea in the middle of one of many Middle East wars, warily watching the deliberately quiet Soviet Navy, the Israelis, and the Arabs. A high-level U.S. Government official in Washington called the Admiral in the Mediterranean with the logical-for-Washington question, "Admiral, are you prepared to fight?" Without hesitation the Admiral of the Sixth Fleet replied, "Yes, sir. Who?"

As should be expected, even the most effective bureaucratic organization can be slow to change how it operates. It took the Navy decades to change from the fixed cannon aboard ship to the self-stabilizing Dahlgren naval guns. The Army used horse cavalry for combat well into World War II. The Air Force, though a premier provider of satellite services, waited decades before establishing a Space Command. The national, state, and local governments have taken many decades to erase discriminatory regulations and laws from their books.

On making a change

But once change has begun, it will continue until stopped. Once an action has been ordered, it will be executed. This response is so ingrained that high-level military officers, when they retire from their service and join private organizations, have great difficulty in understanding that the unstated assumption — an order is a command — no longer holds. Similarly, when civilian executives volunteer for high-level government offices, they are baffled by what they perceive to be endless rules to prevent them, personally, from taking action. In the private sector an order is at most a request for a voluntary response, usually profferred. A handshake is both necessary and sufficient to close the deal, to affirm the commitment.

This fact of life is important for everyone (generals, admirals, CEOs, and employees) to understand, particularly if changes are intended in how the organization is to perform. Because their unstated assumptions about what they can and can't do are often too imbedded for them to be effective, a general probably should not be brought in to straighten things out in a private company, nor should a CEO be brought in to straighten out a government agency.

This fact of life was reinforced by David Packard, long-time chairman of Hewlett-Packard. Packard, after serving the Defense Department for several years as its Deputy Secretary and as an advisor for many more years, brought together top officials from his company and the government and asked that they suggest how each organization might be improved. In effect, should the government be run like a business and/or should a business be run like the government? The unexpected answer was that both organizations were doing well as they were and that neither could offer any major suggestions for improving the other. It was a mutual recognition that each organization operated as well as could be expected in its own domain, rules, and constraints. In the broader context of excellent organizations contemplating a structural change:

> *If a change to or from providing a service is contemplated,*
> *anticipate that a corresponding change to or from a bureaucratic*
> *structure and its rules will be required.*

If the change in organizational structure is too great to consider, reconsider the original change.

Profit seekers: you succeed if you improve *"the bottom line"*

The profit-seeker mode is the structure of choice for *economic* products and services that are billable, measurable, and limited in time, size, numbers, and risk. They are, by far, the largest suppliers of goods to the public and private sectors.

Once a market has been detected, no matter how small, profit-seekers will quickly either develop it and make a profit, or drop it if they can't.

The purpose

Even in the most generous, moral, and dedicated of excellent profit-seeking organizations,[1] profit must be the No. 1 objective. David Packard explained, "Without profit we would not survive to do all the rest," including, among others, customer satisfaction, product quality, compassionate personnel policies, and service to the community.

The profit constraint, nonexistent in service-oriented bureaucracies, can mean that some otherwise valuable products are never produced. For example, at Hewlett-Packard, the profit motive blocked the production of some remarkable instruments and products that would undoubtedly save lives, but only a handful a year. In a sense, after all the other reasons for making a product or providing a service have been discussed and approved, much as they are in government agencies, one more factor is added by the profit-seeker. The product or service also must make a profit.

The Global Positioning System (GPS) story is a good demonstration of that fact. When GPS was started in the early 1970s as a project in the Department of Defense, potential users seemed very scarce. The GPS requires dozens of sophisticated satellites on orbit, specialized control stations on the ground and, in the beginning, expensive and heavy receivers for each user, no matter how small. Army combat troops could hardly carry the equipment. Its antenna made the combat infantryman an easy target. Its economics were poor. Some 20,000 users, operating many hours a day, would be necessary for the GPS to be competitive with other navigational systems such as LORAN, RAYDIST, navigation beacons, light houses, and sextants. At one point, the Air Force (the executive agent for the GPS), on finding that it had too small a customer base to justify continuing its development during difficult budget year, considered dropping it. In any case, not a profit-seeker in the world would have, on its own, considered offering the GPS on the global market. The reasons for this reluctance are instructive.

It took another 20 years for technology to produce small, lightweight, inexpensive receivers and to reduce, somewhat, the billion-dollar investment cost of the satellites, their development, and their continued operation. Then, almost overnight, technology enabled profit-seeking companies to make good profits by producing low-cost receivers. The user market exploded into hundreds of thousands, perhaps millions, of users in all countries all over the world. Fortunately, the original design allowed for nonmilitary users as well as military, or that latent market might not have emerged.

The GPS is now so widely used that it has reached the status of a necessity, generating new private sector markets in airlines, surveying, fire-fighting, deep ocean drilling, trucking, hiking, recreational boating, general aviation, and helicopter rescue — almost anythng that moves and/or needs to know where it is to high accuracy. The likelihood that private industry would have originated the GPS as a service complete with satellites is remote, and certainly was in the 1970s. Among other reasons, it is a one-way service to passive users, like the weather service, that has no way of billing its customers for their use. The achievable profit was limited to selling receivers and plenty of them. But none of that profit returned directly to the government, nor should it.

The GPS, like computers and information networks before it, are good examples of what government research and development (R&D) does best for the profit seekers: establish a small but guaranteed market for a product. To the government, the risk is low in developing a useful if not necessarily economic, product from a research base. But for the profit-seeker, reaching that same point is very high risk because until the utility has been shown there is no way of knowing whether a specific company can make a reasonable profit from it.

The structure

An especially sharp contrast between bureaucracies and profit-seekers is that the former follows strict rules and the latter often rewards its members for breaking them — as long as the result was profitable and not illegal. In addition, lateral (same-level) interaction between organizational units is encouraged in profit-seekers, discouraged if not outright prohibited in bureaucracies.

A further contrast, mentioned earlier, is in incentives for efficiency, cost reduction, and innovative management. The profit incentive is a very strong motivator for self-improvement. Literally, improve or die, particularly if the product has become a commodity, one that many other companies can make, and if one organization is more efficient and can out-sell the competitors' products, then its profits will increase. In the resultant competition, the less efficient producer may be driven out of business. By the same token, going out of business is an expected occurrence for profit-seekers, a rarity for bureaucracies. A degree of job uncertainty, similarly, is expected in the former, not the latter.

Probably the greatest historical test of the consequences of these different incentives occurred during the 70-year period of Communist rule in the Soviet Union. Although the bureaucratic governments of the Soviet Union and the U.S. were comparable in their delivery of noneconomic services, they were vastly different in their deliveries of *economic* goods and billable services. In the case of the latter, the Soviet Union used a bureaucracy and the U.S. used a profit-seeker (free market). Ideologies to one side, the bureaucracy flunked the course on delivery of economic goods and services.

There is another more subtle difference. As mentioned earlier, bureaucracies by their nature are very conservative at the base and relatively liberal at the top. By that is meant that rules are absolute and unquestioned at the bottom, but seen as changeable guidelines at the top. The reverse is true for profit-seekers. Profit margins, generally speaking, are on the order of 5 to 10% which doesn't leave much room for mistakes. Profit-seekers almost never invest in basic research, a 50:50 risk at best. They generally invest in a product development only after a real market has been identified.[2] Schedule overruns in the face of competition can mean major loss in market share. They try hard to make a 20% improvement in time-to-market. Anything less is seen as hardly worth the effort, given the risk of achieving it. The executive level, thus, must be conservative with its resources. In contrast, at the "working level," professionals not only want, they are encouraged to suggest ways to improve processes and products that will increase profits. "Handshake" agreements, instead of detailed negotiated contracts, are treated as expressions of trust and honesty and can be expected to be honored.

Several consequences, therefore, might be anticipated in dealings and issues between government bureaucracies and profit-seeking contractors and suppliers. At the tops of the two organizations, communicating can be difficult and misunderstandings common. At the bottoms of the two, a similar result but in the opposite direction. Only in the middle is there a match between what might be called moderates. No wonder that most of the effective communication, most of the joint committees, and most of the progress and agreements occur there.

On making a change

Given that bureaucracies and profit-seekers are different systems, each doing some things well and some things poorly also leads to a quasi-political pair of heuristics on change:

> *Don't force a business to run like the government* —
> that is, keep regulations to a dull roar.
> and, conversely,
> *Don't try to run the government like a business* —
> but we do! Why?

Organizational cultures: you succeed if you fit in

The Purpose

"Culture" as used here means an *organized* commonality in *behavioral* patterns, *beliefs*, and other products of human work and thought *typical of a group* during a particular period of time.[3] The key words have been emphasized. Cultures can be as large as whole societies, countries, and companies. They can be as small as a partnership, club, or chamber music group. Their purpose can be social, political, ethnic, ethical, charitable, religious, educational, or technical, to name a few. Many are nonprofit, such as universities, business clubs, and honor societies. But not all.

For our purposes here, the most relevant cultures are, for lack of a better term, organizational cultures that strongly affect what excellent organizations can do. First, and most prolific, are management teams composed of a handful of individuals who are interested in a common goal, diverse in their points of view, and personally compatible. These include executive councils, architecting teams, strategic planning groups, task forces, and consulting partnerships. Second are company-wide "ways" of doing business such as the "HP Way," the original Bell Laboratories Approach to R&D, the IBM culture, the Xerox Palo Alto Research Center (Xerox PARC) style, the Marine Corps motto of "Just a Few Good Men," the Honor Code of CalTech students, and so on. They are called organizational cultures to distinguish them from organizations devoted to cultural pursuits, such as museums and social clubs, another subject altogether and outside the scope of this book.

The Structure

Organizational cultures are usually the result of a charter or body of policy, hence the term "organized" in the defining sentence of this section. In each, individual members must fit in or the culture disintegrates and the purpose goes unsatisfied. Diversity of opinion, within the context of the group, is often — but not always — welcomed in the interests of avoiding group think, provincialism, or lack of contact with the real world, so to speak. Some additional cultures of this type include:

- The two different cultures in NASA of manned space flight and robotic space exploration.
- The strongly-oriented quality culture of Japan, echoed in the U.S. in Hewlett-Packard (often described as a "Japanese-style" corporation).
- Almost any military service: Marines, Army, Navy, or Air Force.
- The unique management cultures of General Motors, General Electric, Mercedes Benz, Rolls Royce, Fiat, Microsoft, The Boeing Corporation, and research universities.
- The "objectivity" cultures of The Aerospace Corporation, the Congressional Budget Office, The Mitre Corporation, the Rand Corporation, and selected public media. All are committed to this culture by charter, law, and policy.

Considering their prevelance and long history, one would think that the subject would be noncontroversial. Not so. There is a legal issue whether "fitting in" is exclusionary and, hence, a restraint of trade.[4] Is the culture of an executive council a "glass ceiling" to women and "an old boys club" to the ambitious?[5] Is a design team of specialized professionals exclusionary of the public interest and the media? Should the government "look like" the general public at the executive level,[6] or Boards of Directors reflect the racial and ethnic mix of the local community?[7] And what does cultural diversity mean?

There is little question that the answers to such questions will affect what excellent companies can do. They are highly charged questions laced with hidden agendas, unstated assumptions, and deliberately undisclosed biases.

Confronted with this unpleasant morass, the author looked for answers among those who had faced it directly. One of the most useful individuals was a young software entrepreneur, president of his own small, specialized company.[8] His employees were characteristic of such companies — eclectic in almost everything except making "their" company succeed big time. I had expected as confusing an answer as was the questions: "How can you impose this 'diversity' business when merit is your only hope in this high-pressure, competive game?"

"There really isn't a problem. Without diversity in our company, we wouldn't have had a chance. But the diversity we had and needed was one of perspective, experience, education, and beliefs. I don't know how many times we would get together over some problem and *somebody* would come up with some idea, another would pick it up, a third would bring up a unique experience, and we had it! Had all of us had same perspective, we might as well not have had a group, but just let one individual make all the decisions."

Cross-checked with others, his perspective led to the heuristic:

> *To be successful requires a diversity of perspective,*
> *experience, education, and belief.*

Given that guideline as a management criterion, teams can be formed, executive councils can be evaluated, people can be employed, and progress will occur. Will the choices then correlate with racial, ethnic, gender, age, and other societal groups? Not necessarily. Choosing individuals from these groups because they are members gives no assurance of a useful form of diversity for the specific problem at hand. [TR 88] Indeed, as universities and businesses have observed, such a choice does not even assure that the chosen individuals can or will represent the diverse perspectives of the group they are presumed to represent. Far more important is to assure that the individual can bring a different and useful perspective, experience, education, or belief to the team.

On making a change

The more marked the change from the present business direction, the more likely an organizational change will be needed and, therefore, the greater the diversity of perspective that will be required in architecting and managing the change. Too little diversity and the result will be over-constrained in what might be done. Too much, and the rest of the organization will not understand the intended result and may well block it. All of which makes cultural change in particular more challenging and more potentially dangerous the greater it is. The reader is referred to the previous examples of strong cultures in imagining the problems encountered in attempting to change *internally* from one of them to another, any other. The author can affirm that, for an individual, changing from one culture to another is much harder than changing professions, markets, or sponsors *precisely* because cultures are based on organization-wide, unmeasurable, unstated, "obvious" perspectives, beliefs, and assumptions. Many are the individuals, some high in their organizations, who will say, "But, I'm only a newcomer. I've only been here 10 years."

Cultural factors are built into the minds of top-level executives and their staffs. They, executives and their beliefs, determine what the company does. The remainder of the company, by and large, will perform similar tasks after the change to those before the change. They either know how to make a profit or not. They know how to build a quality automobile or not. They know how to survive in combat, both physically and psychologically, or not.

A dramatic change in direction, to be successful, may require the termination of the top managers, not because they are incompetent but because they are *too* competent in what they do best and will keep trying to do it — like the six eagles who could not change.

Unfortunately, that is not the end of the story for the company. The new managers and executives, presumably expert in the new direction, do not know from whence to start. They do not know the people, the reasons the people do what they do, and consequently how to communicate with them efficiently. Or at least for some period of time.

Therefore, even after an abrupt change of direction is decided, its implementation will take time and may have to be constrained in scope and extended in time. Some applicable insights:

> *When implementing a change, keep some elements constant to*
> *provide an anchor point for people to cling to.*
> (Jeffrey H. Schmidt, 1993)

> *Unless constrained, change has a natural tendency*
> *to proceed unchecked until it results in a substantial*
> *transformation of the system.* [SO 93]

> *Given a change, if the anticipated actions don't occur, then there*
> *is probably an invisible barrier to be identified and overcome.*
> (Susan Ruth, 1993)

Matching structures and products

It should not be surprising that what an organization produces and how the organization is structured should be at least related to each other. A football team is organized according to the plays intended. In point of fact, the defensive lineup is allowed to change right up to the moment each play begins. An orchestra may change with every piece played at the concert, sometimes getting bigger, sometimes smaller, sometimes to the complete exclusion of a whole section, and always with the addition of a soloist. An architecting team changes its members as the product line is developed.

Of the three categories of organizations given here, the profit-seeker characteristically is the one that will most closely tailor itself to what it does. In addition to the classic hierarchy, profit-seekers may structure themselves into matrices, pyramids, segmented rings, projects, and even phantom or virtual configurations.

Matrices are usually chosen when many products (or projects) exist, each calling for a wide array of disciplines. It is called a matrix because it resembles a grid with control coming not only from the top (say, the disciplines)

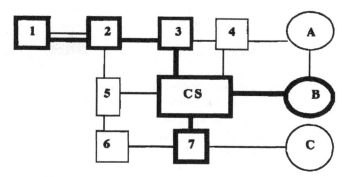

Figure 6.1 Overlay or phantom network (1, 2, 3, 7, and B) with alternate backup routes (through 4, 5, 6, and A). Note: Normal routing through main network (heavy line); alternate routing through main network (thin line); most vulnerable transfer node: 2.

but from the sides (say, the projects). It is a structure that permits fairly rapid, incremental change from one project to another because the key professionals already are in place in the discipline divisions. To them, adding or deleting a project is not a problem. However, a matrix becomes ungainly if most of its projects begin to fall away. The principal disadvantage of the matrix is that every professional (say, a propulsion expert) works for at least two bosses, his division, and several project directors.

Pyramids, as the name implies, are built from relatively autonomous levels, each solving its own problems and each whole level reporting by many ties to levels above and below it. Its difficulty is that it is not clear who is in charge of each level and, hence, who can make a change.

Phantom or virtual organizations have no physical reality on their own. They can be said to "overlay" the parent organization, using but not being aware of which interconnecting lines and nodes provide the routing. A good example is the "personalized" telecommunications net of Figure 6.1 which behaves to its customers as if its components were completely devoted to its service, though in reality it is a shared network with many others able to use the same facilities. This structural form is characteristic of Internet in which connections physically can be made on demand between two or more subscribers at will, unlike the hierarchy in which connections are fixed.[9]

It is not necessary to go into these structures or others in detail to make the point; each will affect the using corporation differently. The change from a structure that has served only an American market to one that must also serve the European and Asian markets will affect and be affected by the organization's structure. A responsive change should obviously affect marketing and sales, but what about manufacture, design, and research. Should the new organization be partitioned by region, the production waterfall, or technical discipline, for example? Conversely, which partitioning would best serve each or all markets? And how should the partitioning be done? An earlier heuristic suggests:

*Partition for the minimum necessary communications
between the elements.*

So much for the easy part. But suppose the partitioning is different for manufacturing than for marketing or research, which one should be implemented? Part of the answer is supplied by the heuristic itself, the organization's communication system. Distance, proximity, quality, and quantity are no longer serious obstacles. It is now common to have, say, a software division with its sections in the U.S., India, England, and Canada operate as if everyone were in the same building, only on different floors. Indeed, interfloor communications is often more difficult to implement than intercontinental, and the boss can be anywhere in the world at any time. If desired, there needn't even be a physical headquarters, a real puzzle for national regulatory bodies, a problem highlighted in Chapter 4.

Stable intermediate structures as milestones in change

Clearly, if major change in products and processes is to be made, then major changes in people, policies, and procedures will have to happen as well. In architectural terms, the organization will have to be "re-architected," the degree of which depends upon the change desired. There is an important heuristic, a derivative of one from the software community in the 1960s, that states:

*A large change is best made through a series of smaller, planned,
stable intermediate states.*

It suggests that the end structure be achieved through a set of well-designed intermediate steps, each one stable enough that the progression could be stopped, indefinitely if desired, at known, predetermined points. Metaphorically, it suggests that when hiking over dangerous territory, steps are better than uncontrolled slides down potentially slippery slopes to well past the intended destination.

A note on promotions and transfers

The choice of structure in an organization does more than control its operations and influence the design of its products. It also affects the promotion possibilities for its people and, hence, what an organization, particularly an excellent one with exceptionally talented people, can do.

A hierarchy, whether in a bureaucracy or not, is a relatively stiff structure. Reports go one way, control the other. The implication is that the same holds true for the people in it. They stick with their specialty, gradually broaden it, and get promoted "up the line" when the right opportunity occurs. Particularly in administrative hierarchies this situation tends to lock professionals into narrow specialties (personnel, facilities, legal, contracts, etc.) and, hence, block them from higher, inherently multidisciplinary, promotions. It takes higher authority to take the risk of promoting a facilities person to oversee legal and the other specializations.

The matrix is structurally better for the frustrated professional. At least there are opportunities for lateral moves, more working relationships, and, consequently, improved visibility. But the best chances are through company policies that recognize if and when it is in the system's interest. When that is true, then it is in the organization's *and the system's* interest for many of the best people to move laterally across interfaces and to understand the imperatives of different specializations. Given a policy that encourages voluntary lateral moves — not easy when the customer wants the world's best expert to stay in place "forever" — the result is a far better qualified *systems* organization, a greater pool for every promotion, and a lesser chance of a first-rate professional being "stuck" in position by being too skilled to be moved or delayed by the longevity in place of a well-qualified superior.

Summary

Different structures with necessarily different rules affect the behavior and performance of organizations much as they affect most other complex systems. Some structures are better suited than others for the objectives that are sought; bureaucracies are better for noneconomic service, profit-seekers for economic products, and cultures for dedicated causes. Important for companies contemplating change, is that different organizational structures can either aid or preclude the ease with which change can be made. Heuristic guidelines are suggested for the decisions that need to be made.

Notes

1. The author, having worked for 4 years at the headquarters level, uses Hewlett-Packard as a preferred model of such an organization.
2. One of the most effective actions a government can take is to guarantee a market, albeit small, for new, higher-risk products and processes. Computers and microchip production are two of many examples. Once guaranteed a market, profit-seekers are willing to take the lower risk of achieving a quick market expansion.
3. Notice that this definition makes no mention of ethnicity, race, gender, or other unavoidably physical attributes of an individual or grouping. It is one of behavior, beliefs, and work patterns. In short, this definition is mental, not physical. It closely resembles that in *Webster's II, New Riverside University Dictionary*, 1984, p. 335.
4. Yes, according to a judge who decided that codes of ethics for collegiate coaches excluded every one who didn't choose to abide by them as a restraint of trade and, hence, illegal.
5. Yes, according to a number of women's advocacy groups.
6. Yes, according to the Congress in demanding that almost all National Research Council committee deliberations and meetings be open to the public and media. This demand has since been modified to be less draconian.

7. Yes, according to the administration of President Clinton whose appointments to national office are intended to "look like the country."
8. Benjamin J. Bauermeister, president of Elsewhere, Inc. of Seattle, WA, since merged into Hewlett-Packard.
9. For a more detailed discussion and figures, please see [RE 91 274-80].

chapter seven

Decision making in complex organizations

Introduction

As has been mentioned earlier, organizations can be viewed from many perspectives. In this chapter they are seen as the almost irreversible consequences of assumptions and decisions made in the past. Few of the key decisions can be reversed without affecting decisions that came after them or without negating some that preceded them. Some, indeed, may have been forgotten, many are simply unstated, taken for granted.

For example, in the late 1940s the management of the Jet Propulsion Laboratory (JPL), a CalTech-affiliated, private research laboratory, faced a key decision, whether to change from further developing Jet Assisted Take Off (JATO) aircraft or to divest that work and begin research on guided missiles. The JATO course would lead to what became a commercial rocket-making company, the Aerojet Corporation. The guided missile course would eventually lead to a NASA-sponsored, nonprofit center for the exploration of the solar system — though none of the latter course was known at the time of the decision. Once made, there could be no turning back, either for JPL or for Aerojet.[1]

It is a common and erroneous assumption that decisions like these are made one at a time, based on well-structured, well-understood issues. Not so. Decisions on technology, management, staffing, core competencies, facilities, future clients, location, and even the continuation of the ties to CalTech professors had to be considered if not made on the spot. And further down the road, each of these collateral decisions would be different depending on which of the two courses were followed.

The decisions to change the course of JPL *defined* that organization then and for at least the next 50 years, as they did Aerojet. It is not a unique story.

Similar stories can be told about Boeing and McDonnell Douglas, about Rockwell International and Collins Radio, about Thompson Products and Ramo-Wooldridge.

At a much smaller scale a story can be told about implementing affirmative action[2] by making coordinated policy changes in personnel practices, legal actions, retirement programs, staff transfers, promotion procedures, executive contracts, awards programs, parking places, position descriptions, technical staff requirements, nepotism, and conflict-of-interest rules — a dozen different defining elements from which emerged an effective affirmative action culture for the organization as a whole. And it was done without the use of mirrors or preferences.

Clearly, decision making in a complex organizations is not, and cannot be, a matter of one decision at a time dependent upon which issues happened to show up in the chief executive's office on a random Monday morning. Decisions, instead, are made in sets and in a way that keeps the organization in some sort of balance before, during, and after the change.

The subject of decision making in sets will be followed by showing the importance of the *order* of decision making in determining the final outcome, a subject that deserves special attention when making radical changes in direction at full speed.

This chapter ends by re-addressing another aspect of decision making in complex organizations that strongly determines what they are, why they are, and where they might go or not go; namely, making sound decisions when insufficient information is available to prove them sound.

The format, in a possibly welcome change of pace for the reader, is deliberately more graphic than in the foregoing chapters. Somehow, decisions and their rationales are more easily portrayed graphically.

Inherent tensions in complex organizations

Visualizing tensions

Managing has always been a matter of delicate balancing of tensions brought about by conflicting interests. In fact, a good argument can be made that balancing these tensions is what makes organizations work. They help assure that no major factor or element will be inadvertantly ignored, that no good idea will be left without a sponsor, or that no legitimate perspective will be forgotten.

With a respectful nod to bicyclists who might be surprised at the idea, an organization can be usefully compared with a bicycle wheel. The bicycle wheel is one of mankind's most remarkable inventions. It is the key element in the most efficient human-powered transport system ever built. Instead of using heavy spokes and a rigid outer rim, the bicycle wheel uses almost wire-thin spokes in tension. The mechanical engineer would call this a "prestressed" structure in which a compressive load is accommodated by a backing off of the tensions in the spokes. If there isn't enough tension and the imposed load exceeds the tension prestress, then, of course, the wire-like spokes abruptly bend and collapse. In organizations, if an external force overloads an interface and the elements next to it, the resultant shock wave

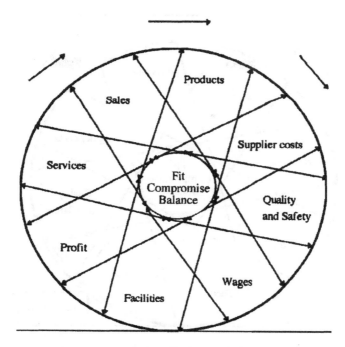

Figure 7.1 The bicycle wheel metaphor for complex organizations.

may demolish the entire structure. At the same time, if the tensions are too great, they can so warp the structure that it cannot perform properly.

The bicycle wheel metaphor for complex organizations

Figure 7.1 is a sketch of the rim-and-spoke structure of a bicycle wheel rolling along from left to right, progressively cycling through facilities, wages, products, and so on through profit and then around again. Management of this rotation is accomplished at the hub by fit, balance, and compromise. The rim represents the interface or boundary with the world "outside," and especially with the ground underneath which progressively supports each part of the wheel as it rolls (merrily?) along.

Figure 7.2 shows the wheel just as it meets an obstacle in the road; that is, just as the organization is about to be subjected to a sharp external force, an event. For the mimicked organization at this instant, the force shows up in wages (a union demand, a sudden inflation, a competitor's unheralded change) and sends shocks along the rim and up the spokes beginning with safety, quality, and facilities and propagating throughout wheel as it is forced up and then dropped down with a jolt on quality. A mechanics professor would no doubt show how the opposite side of the wheel "mirrors" each of these jolts and how the changing tensions could add a self-induced bounce or two.

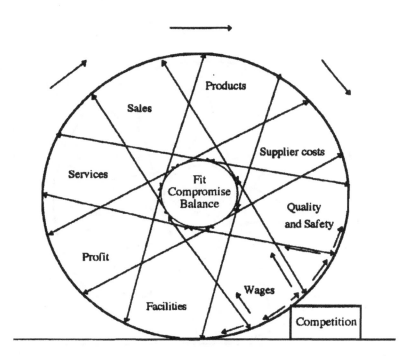

Figure 7.2   Rocks in the road.

The graphic lesson here is that a complex organization, like a bicycle wheel, responds to an external force as a dynamic whole, maintaining its integrity and capability through internally transferring and absorbing the shocks imposed upon it.

Balancing tensions between performance, schedule, cost, risk, facts — and perceptions

Figures 7.3 through 7.6 make a further point about decision sets. Figures 7.3 begins the sequence by sketching out one encountered routinely in managing product line architectures: the balance among performance, schedule, cost, and risk, with "architecting" shown in the middle as a balancing process. Figure 7.4 sketches in what might be called the second echelon; that is, the sources that underlie performance, schedule, and so forth. Figure 7.5 then adds yet another tension, one particularly important in organizational decisions, that between facts and perceptions. The difficulty is that only one end of the axis is rational, yet both are equally real. The reader may remember one of perception's more blunt but unquantifiable heuristics:

"I hate it!" is direction. (Gradous, Lori 93)

The underlying reality of this tension pair is shown in Figure 7.6. Clearly, experience, associations, and presentations have no place on a numbers scale,

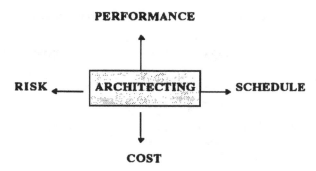

Figure 7.3 Four common product or service tensions: first echelon. (From Rechtin, E. and Maier, M.W., *The Art of Systems Architecting*, CRC Press, Boca Raton, 1997. With permission.)

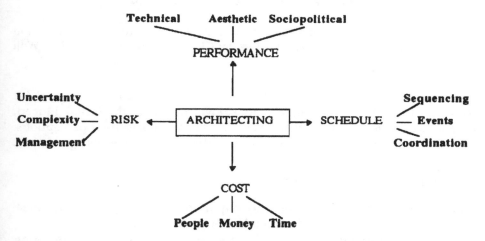

Figure 7.4 Underlying sources of the four tensions: second echelon. (From Rechtin, E. and Maier, M.W., *The Art of Systems Architecting*, CRC Press, Boca Raton, 1997. With permission.)

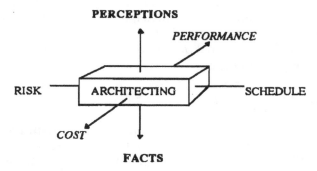

Figure 7.5 Adding facts vs. perceptions: first echelon. (From Rechtin, E. and Maier, M.W., *The Art of Systems Architecting*, CRC Press, Boca Raton, 1997. With permission.)

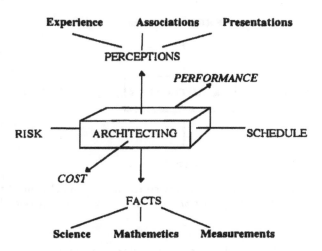

Figure 7.6 Sources of facts and perceptions: second echelon. (From Rechtin, E. and Maier, M.W., *The Art of Systems Architecting*, CRC Press, Boca Raton, 1997. With permission.)

but a manager would be daft to dismiss it as "unscientific and illogical." It is very real and, at times, much more powerful than "the facts." Indeed, fellow architects and engineeers, facts can change faster than perceptions. When they do, we call them scientific breakthroughs and *everything* changes.

Decision sets and emerging capabilities

Making decisions in sets is especially important when they affect the emergent system capabilities of the organization as a whole.

Emergent capabilities, as has been seen, are the result of interacting relationships among organizational units. Therefore, decisions affecting these relationships will by definition affect the emergent capabilities of the organization, enhancing, deleting, or precluding an organization's capability from emerging. For example, a decision to buy an off-the-shelf software component, logical enough for one management unit, may block a future option for inventory control upgrading in another. The same decision also might result in staff changes that derail the company's plans or assumptions to build new core capabilities.

Decision making in sets

Decision making in sets, for all the time it takes to put all the pieces into rough balance before acting, may well save time in the long run by avoiding endless reiterations until balance and stability are regained. In the bicycle metaphor, the decisions to be made are whether to increase the tension in a given spoke or spokes. They help avoid making one decision that calls for another which calls for an adjustment for the first one, and so on, back and forth until the pair are consistent. Then comes a third decision, calling for still more adjustments of the first two.

Such was the case in a several billion dollar construction project. Great presssure was placed on the project to be finished in an almost impossibly short time. Construction was begun before the design was complete. Hence, change orders were unavoidable. Before long, change orders to change orders began to appear. At one stage there were 1000 of them a month, some of which applied to changes made less than a month earlier, which no doubt affected still others before and since. It was a change controller's nightmare. Then, about half way through the job, an external, not-unexpected, event caused a 3-year postponement of the project's mandated completion date. The construction project began to implode in exasperation, and finally the half-completed structure was mothballed, put in storage, so to speak, until "next time." It wasn't the first time that the lament was heard,

> *Why is there never enough time to do the job right,*
> *but always enough time to do it over?*

The answer depends upon the circumstances, of course. In this case, a number of uncoordinated decisions contributed — a decision to proceed with construction before design completion, a series of decisions to change without consideration of the consequence to other decisions, and an inability to consider decisions in sets or "blocks."

On the formation of sets

The next question is which decisions affect which capabilities and vice versa? A number of emergent organizational capabilities have been mentioned in preceding chapters. Among them are excellence, success, objectivity, profit, and diversity of perspective among them. None of them can be achieved, much less all of them, by any single decision. Some decisions, by their nature, will have to be made in parallel, others sequentially, others sooner or later. Some may be overtaken by events, others pushed aggressively. How to proceed?

Fortunately, this is not a new problem nor is the solution new. The decision process suggested here is similar to the one used by architects for generations. Ask questions. Learn. More questions. Scope, and so on. It will be developed further in Part IV, but for the purposes here it will be short-formed by example.

The first step is notional; that is, to have some notion of one's objectives at the highest level of generality that is useful, then begin scoping or "scoping" the applicable ones.

As an example, let's begin with profitability as the general objective. The next step is to bound or scope that desire by context and tolerance for risk. As an illustration, assume that the context and risk are stated as smart systems, high risk, and high profitability. Which, of course lead to still more questions. Product size and quantity? How soon? Core capabilities required? And the like. If a quite new product is involved then the product must first be synthesized, conceptualized, engineered, produced, and sold, a collateral *set* of decisions need to be made early or they could mean over-constrained decisions later on.

To continue the example with a technical fact of life: a smart system requires both hardware and software. Does the company have the necessary capabilities to develop each or must one of them be further developed or acquired soon and, hence, expensively? Unfortunately, the cultures, processes, products, and management of the hardware and software are different — sometimes incompatible — and, consequently, may be too inconsistent with the culture and management style of the company to be worth the upheaval. If so, then another set of decisions has to be made and iterated with the original ones before entering the smart systems business even for minimal profit and for some time. One possible decision is to go back and start over…

Making decisions in sets can be uncomfortable. Some of the organizational units will want the decisions affecting them made as fast as possible. Perhaps they are about to "cut metal" or to commit to a place in the waiting line of an important supplier.[3] Others will want them delayed, anticipating developments which may make a particular decision either obvious or moot. Few of the units will know or understand the needs of the others. The decision maker is in the middle.

On the order of making decisions

Even within the limited set of critical decisions, the *order* in which they are first considered can drastically affect the outcome. It is not difficult to see why. Imagine that there are an equal number (say, 1, 2, 5, or 10) of turn left and turn right decisions to be made in driving a car in a gridwork of streets. If all the left turns are made first, followed by all the right turns, by the end choice the driver will have remained within a few blocks of the starting point, getting almost nowhere. If the lefts and rights are made alternately, the driver will have dog-legged in one of four widely separated, diagonal directions depending on the initial direction and the first decision. This example may be trivial, but the conclusion is not:

In decision making, order counts!

At the simplest level, is a building design begun from the inside out or from the outside in? Is a space program defined from the top down (the mission of the spacecraft) or from the bottom up (launch vehicle and facilities)? At a management level, should the manager of a project be appointed before or after the architecture is conceived? At the corporate level, should the chief executive be chosen before or after the board of directors has decided on the direction the company should take? In forming a team, in what order should the members be chosen — in order of rank, competence, or mission priorities?

The author recalls a time when as a middle manager he badly needed a particular decision. It didn't seem to be forthcoming in a timely fashion, considering its importance. After an impassioned advocacy by the author, one of the several wise managers for whom the author worked[4] said, "Eb, I'm quite sure you are right, but there are other decisions which have to be made first, which then can make the one you need straightforward and reasonable. But if the decision you need were made now it wouldn't have the foundations which you have assumed but which don't exist yet." It was one of the most pointed and appropriate pieces of management wisdom in the author's career. He was right. The future was as he predicted.

The point that was made, that earlier decisions were necessary to support a later one, should be paired with the counterpoint that many decisions also can *constrain* those that follow them. Indeed, unless care is taken in the early decisions, the result may be so over-constraining as to make further options impractical if not impossible; a true dead end.

Premature decision making

One of the most common examples of making decisions in the wrong order is deciding on "the solution" before deciding on "the problem." The result is the familiar "solution looking for a problem." Such a designation was probably appropriate, in the author's reasonably informed view, for very high-power lasers, today's space stations, and most strategic (nuclear armed) bombers after CY 2000.[5] But this personal view is certainly easier in retrospect than in prediction.

A solution may yet find the problem just in time! The widely used Global System, the medical laser, the hand-held calculator, and even geostationary satellites were very far along in development based on arguable applications before unimagined latent markets appeared, making all of them successful and profitable. The electric car may reach this point near the turn of the century. It began years ago as a proposed solution to a real problem, atmospheric pollution. But development was begun and legislation enacted (and effectively reversed) before a practical energy storage mechanism had been developed that could be competitive with that of chemical fuels. In the meantime, other so-called hybrid possibilities — using chemical fuels much more efficiently — have arisen which may, for all practical purposes, solve the environmental problem sooner at much less cost. Or, it may simply shift the environmental costs into other domains (nuclear, fuel-efficient remote power plants, renewable fuels, social costs, and so on) when and if oil prices again start climbing.

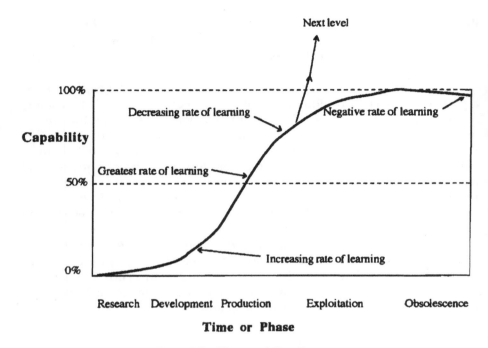

Figure 7.7 The capability S-curve.

Belated decision making

The opposite of premature decision making is belated decision making, making decisions too late. The six eagles problem, in other words.

Whether a decision is premature or belated depends upon the decision maker's judgment of where a product, or management technique, or research program, is on its "S curve" (Figure 7.7). Leaving it too soon risks missing its most profitable payoff. Leaving it too late risks powerful competitors entering the marketplace with better architectures, technologies, or insights.

A mathematician, strategic planner, or experienced manager can see the issue involved relatively easily from looking at the S curve. At its beginning, the rate of change (the slope of the curve) is low, corresponding to a research and feasibility stage. The slope gets steeper and steeper and then begins to flatten out until very little is changing at the far right end. The product has become commonplace, producible — or reproducible — by anyone. A more successful strategy would be to plan at least one phase ahead throughout the project lifetime. An obsolescence or exit strategy should be planned, by this reasoning just when the slope is greatest; that is just when the greatest progress is being made. No question but that this will be hard to do in the enthusiasm of that time.

A more aggressive strategy, arguably, is to think about leaving the S-curve and using its upward momentum to reach the next higher one, whether it be in diversification, sophistication, breadth of product lines, or merging with others. In day-to-day terms, if the lessons learned per day are decreasing instead of increasing, develop an option of moving onward and upward. By analogy to launch rocketry, which the curve also resembles, plan ahead on "staging." Each stage leaves the earlier one when the latter is going as fast as it can.

Briefly then, plan ahead when things are going very well, not after an unmistakable observation that they have gone stale. A belated decision will cost time, if not money and people, because change takes time. More time can only be bought before the need and never thereafter. Thus, because it may be some time along the S-curve before a decision to change can actually occur, it is none too early to plan obsolescence when none is yet in sight.

These lessons deserve attention at the executive level in the organization because they relate not only to the company's future but to that of its people — who also must plan for it. This career planning is especially critical for members of architecting and other teams. Ideally, architects would remain for the duration of their project; that is, from conception through certification of readiness for use. In practice, that may be difficult both for the architect and for the architects' employer. Yet, it is better to have a plan in place when the moment of decision arrives than to have none at all.

Some of the most important sets of decisions may have to wait for key events. For example, while there might be a good window of opportunity to move from one market or capability to another, there also must be an opening to enter. To make the move, two windows have to be open, not just one. For that reason:

> *Don't leave until you know where and when you are going!*
> And, of course,
> *Know why you are leaving* — so you never have
> to look back and ask, "What if..."

The obvious rejoinder to this admittedly idealistic strategy is that much of the information needed to carry it out cleanly and well is at best hazy and, at worst, nonexistent or denied.

Decision making with insufficient information

And finally, decisions in a complex organization will have to be made lacking information critical to its future. Some information won't exist until too late. Cost modeling in particular will necessarily be imprecise, though it might not look like it in a six-figure printout! Some will be denied by competitors. Some, especially empirical data, may be hopelessly time-consuming and expensive to collect. The strategy of waiting until more information is available may lose the game in the meantime. For example, no single piece of information in the past 50 years was more critical militarily, politically, and economically than knowing if and when the Soviet Union might break apart or the U.S. fall into decline, for that matter. Yet companies had to decide whether to bid on weapon systems, to embrace or reject China and Cuba, or to include the Warsaw Pact countries as potential markets. At a lesser scale, individuals had to decide what languages to learn, what professions to choose, and what companies to join.

In fact, decision making in complex organizations is seldom based on full and complete information necessary for a well-informed, thoroughly justified, inspection-proofed decision. Because not taking any action or waiting for more information are also decisions, there is a fairly high probability that any decision will be only partially right and, retrospectively viewed, may prove to be very wrong in the future. The eagles waited too long to abandon an obsolescent architecture. The electric car manufacturers came to market before the critical technologies were ready. President Clinton proposed national healthcare before the country was ready for that much of a change. What should be done?

Of course, it depends upon the circumstances, but a heuristic called "Choose. Watch. Choose." is at least a start. It states that, given a choice:

> *Choose as best you can.*
> *Watch to see whether solutions show up faster than problems.*
> *If so, the choice was probably a good one.*
> *But, if problems are showing up faster than solutions,*
> *revisit the decision that caused this to happen and choose again.*

This heuristic is displayed graphically in Figure 7.8 which says keep going and in Figure 7.9, which says to go back to the "Square one" decision. What does "solutions before problems" mean in these heuristics? It means that when problems keep coming up, as they always do, the solution is apparent without much further thought. It means that in a presentation, when an unexpected question comes from the audience, the answer seems

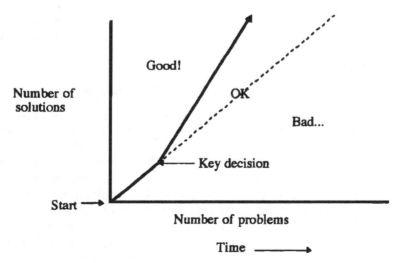

Figure 7.8 Keep going!

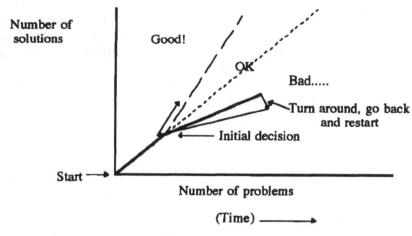

Figure 7.9 Restart! Reboot! Go!

to spring up of its own, though the presenter never considered the problem before. In governmentese, it means that the presenter can talk "off the chart" into unexplored territory without concern. It means that the basic strategy, the conceptual architecture, and the management approach are consistent and successful, so far. In contrast, when problems come faster and faster, with each one a frantic fire drill, when things are generally going to hell, when presenters "have to get back to you with that," go back to Square one. The project is in deep trouble.

Going back to Square one is difficult, time consuming, often expensive and dangerous to one's professional and political health. But it is better from the project's perspective than continuing to a final disaster. If arriving at a decision (that leads to solutions exceeding problems) takes skill and luck, then reversing course when problems overwhelm solutions, takes courage and leadership.

The power behind this particular heuristic is in the words, "watch for." Watching and waiting is not the game. It is watching for a particular situation to arise. Better yet, to begin to arise. Somehow, experienced managers and professionals alike sense this situation remarkably early, sometimes within a subsequent problem or two. The situation just doesn't play, or feel, or smell, or taste right, depending on how the individual describes the feeling. It is very real and has served many people well. But no one, this author included, would guarantee that it always works. Sometimes, just over the difficult hill is nirvana. Just don't bank on it.

Summary

Because complex organizations and systems are indeed complex, so must be the decision process. Just as complexity involves multiple, simultaneous interactions, decision making involves multiple, simultaneous problem resolutions.

The management problem, therefore, is to aggregate closely related issues and to treat them as collectively as possible. Not the least of these is to assure that the desired emergent properties of the organization as a whole remain intact. Some of the more critical aspects of such decision making are described, including inherent tensions, emergent properties, sets of decisions, the order of decisions, and decisions based on incomplete information. Comparable decisions and actions of the contributing professionals are indicated, based on the same lessons learned.

Notes

1. For another, well-documented decision history see J. E. Steiner [STE 83] on Boeing commercial aircraft.
2. At The Aerospace Corporation, 1977-87.
3. One of the longer waiting lines is for large metal forgings, a line four years long and only a single supplier. This situation is likely to worsen as more second tier suppliers leave the sharply-decreased defense business.
4. ex-General Al Luekecke, then Deputy Director of the NASA/CalTech Jet Propulsion Laboratory, circa 1961.
5. In contrast to low powered lasers (a few kilowatts), high powered (megawatt) lasers have yet to show any utility for space, air, land or water defense. They are much too heavy per energy delivered to a remote target. Space stations are being built with no clear purpose. Earlier scientific, military, medical, and entertainment possibilites have vanished, leaving only the near-purposeless human colonization of Mars, for which the earth satellites are backward steps. Given the political to use nuclear weapons, ballistic missiles are far more effective than nuclear bombers for all but moving targets, few of which will exist or be detected by bombers flying into a nuclear-devastated countryside.

Part IV

The foundations of organizational architecting

On creating insights

The world of the lawyer and public servant is
one of rules and precedents.
The world of an artist is emotions and perceptions.
The world of an engineer is one of facts and figures,
science and mathematics.
and
The world of an architect is one of ideas, insights, and inspiration.

Introduction

Any methodology has a built-in challenge to it. The challenge to rules is completeness — there are never enough. The challenge to engineering is the correctness of specifications — they are never perfect. And the challenge to insights is their credibility — there are no mathematical proofs that they will always work. This last challenge, for better or worse, cannot be answered by replicable measurements, a fact shown in Chapter 1. Credibility can only be answered by examples and results; that is, if an insight has been shown to work before, it's probably okay to use now.

The next chapters will lay out the context and use of insights. But why should they be believed as more than one man's opinions? To respond to this question, this introduction will show the results of 10 years of instruction, challenge, and use of architecting insights — heuristics and metaphors — with results to date. Examples have been introduced throughout the text. Suggestions for use in architecting organizations will be given in Chapters 9 to 11. But the real answer lies in the reader's perception of how useful it can be for the problems at hand.

If you would learn, teach

More than 10 years ago, when the author began his exploration into the unknowns of systems architecting, it was not clear whether the concepts of civil architecture and engineering were transferable to the later branches of engineering (mechanical, chemical, electrical, and so on) to the more recent information and systems branches.

Nor was it clear how the subject should be taught, an imperative for any true profession.[1] In fact, there were neither courses, textbooks, or degrees. The teaching method turned out to be novel but straightforward as described in Chapter 8. It was validated when it was readily used by the succeeding professor to teach hundreds more students.

If the original author–professor thought of himself as the guru of the field, he was soon disabused by his students. To enroll in the subject, students had to be full-time professionals with at least 3 years experience in the systems field. Most were working in major defense programs, though there were notable exceptions. The average level of experience was 7 years, the maximum 15. The author was older, but the students were closer to the firing line and showed it. They knew what the author did not — software, the latest technologies, and the transition from Cold War to commercial competition — and had already learned when and how to be skeptical of anyone that claimed to be smarter than they!

The professor would profess some insight or other. The students would take it "home" to their jobs, try it out, and report back. Later they would discover insights from their experience in their own projects and bring them to class. Speak of testing a hypotheses in the real world, these systems insights were tested within weeks.

The first hurdle

As the author discovered very early, systems architecting is both an applied science, like engineering, and an art* like major parts of (civil) architecture, law, and medicine. The applied science of systems engineering, backed by a Bachelor's degree in an engineering field, is well taught elsewhere. So, within a few weeks of the first evening class, it largely disappeared in favor of as much time teaching and learning the art as possible.

Texts on the theory of (civil) architecture by Jon Lang [LA 87] and P.G. Rowe [RO 87] were studied in the first year of classes, resulting in an early decision to concentrate on the "heuristic" method, rather than on either the normative or rational ones.[2] There were several reasons for this concentration. It was unique in focusing on nonscience, nonmathematics, problems not taught anywhere else, and it used tools not based on replicable measurements. It demonstrably taught more about the practice of architecting and how it "felt" than any other subject. While learning about other theories might be worthwhile, the unique insights of the heuristic approach provided more immediately useful results for problems not otherwise solvable.

But there was a hurdle to overcome. All the students, and indeed the professor, himself, had been brought up in the scientific method and, as

---

* "Art" is used here in the sense of "practice" or profession based on conceptual skill and intuition.

Chapter 1 indicated, the art of systems architecting was not only not science, it could proclaim no provable "truth." Some of the engineering students never overcame that hurdle but tried to the best of their ability to accommodate. Generally speaking, their approach was to add as much systems architecting to systems engineering as they felt they could use. The great majority of the students welcomed the focus on the art because they had faced just such unsolvable problems, but had no tools with which to treat them.

The student profile

Although most students were engineers, some also were engineering managers and project leaders. About 10% were women or minorities. There were few foreign students, no doubt because a course prerequisite was fluency in English, almost to the point of high philosophy — and there were some astute philosophers in the course! Some students were taking other courses at USC, a number which steadily increased as the course developed into a multicourse Master's degree.

The pioneers

About one in ten students, given the opportunity, could become very good, even world-class systems architects. Most, the pioneers, have already made fundamental contributions to the field. These certainly include Ken Cureton, who wrote the rules on how to select and create insights and architectural teams [CU 91 and 92]; Kathrin Kjos, who extended systems architecting to safety designing; Jonathan Losk and Tom Pieronek, who profiled systems architects [LO 89 & 90] and [PI 90]; Ray Madachy, who showed how to use Hypertext for browsing the expanding fields of systems architecting heuristics [MA 91]; Archie W. Mills, Jr., who explored architectural innovation [MI 91]; Mark W. Maier [MAI 95] and [R&M 97] who ranks as a founder of systems software architecting; Jerry M. Olivieri [OL 91 & 92], who laid out the differences between systems and biological architecting; and Marilee J. Wheaton, who demonstrated the close relationship of cost modeling to systems architecting[WH 89 & 90]. Before the first class session in 1988, the author knew nothing of any of these aspects of the field. Verily, if ye would learn, teach!

About 6 in 10 students (the converts) appear likely to use or exploit systems architecting ideas and insights as supervisors, project managers, and program directors. The remaining 3 in 10 no doubt got the message, so to speak, but probably couldn't restate it in their own words. Less than 5% "left early, no record" as is permitted at USC. The course had been designed to be a tough one and it has retained that reputation ever since.

Ahh, women

When, after 6 years of teaching — but admittedly based only on a small sample — it began to appear that the women engineering students, proportional to their numbers, were more proficient in the art of architecting than the men, the puzzled professor asked other women, including his wife and her friends, of why that observation might be true. The response from women in general was, "Of course! We are used to dealing with such problems all the time. You men (engineers) always want to get everything neatly spelled out with only one correct answer. We know all about problems that never go away and about just trying to get problems solved any way we can. Call it women's intuition at work." So much for *that* question. If it and its reason are true, this observation indicates a potentially important resource for the engineering and architecting field — and a unique opportunity for women in this field.

Student impressions and report backs

The unsatisfied engineer

If there was ever a common comment from the students on first experiencing systems architecting it was, "Gee, when I began as an engineer, I thought that this was what engineering was all about!" To the author this was a particularly poignant remark because these students evidently had not had the opportunity for being really creative as engineers that the author had, and because of that, still revels in the field. Perhaps the systems engineering paradigm of choosing among (given) alternatives using mathematical and algorithmic tools to get the "optimum" single right answer is stifling creativity and its exhilaration. Perhaps taking predetermined requirements and "flowing them down" into work statements has become just a routine, making engineers appear to be, well, dull. As one engineer grumbled, "There aren't enough Renaissance types around anymore." Perhaps...who knows?

The "if only" lament

The second most common comment from students and CEOs was the lament, "If only we had known this when we started..." Before it became apparent that insights were effective tools and approaches, the problems that they treated were put aside as "too hard" or "that soft, squishy stuff," remaining as unobserved as underground streams until they eroded the foundations and rationale of the projects themselves to the point of collapse... And so those problems remained quiet and as unobserved as underground streams until they eroded the foundations and rationale of the projects themselves.

That unseen erosions, thankfully, were recognized by the students involved in such projects, were reported to their supervision, and, once articulated, could be addressed. The welcome results were duly reported back to the course professor.

Most student–professionals talked enthusiastically with their supervisors about what they were learning and doing. Their comments were greeted with interest and approval for learning something of immediate use in their work and for writing reports that the supervisors could use to good effect. More than one project supervisor, glad to see where the troubling elements really were in a project (and what to do about them) revised or cancelled tasks to free the professionals to work on better ones. Reported troublesomes were lack of clear purpose, unscaleable research, undefined success, likely competitor reaction, lack of bailout options, and misunderstood client nervousness.

A successful eagle

The most remarkable story was of a program manager who, as a student in the course, talked with his company's defense department sponsor in the language of systems architecting to the latter's enthusiastic acceptance, who then ordered all in his command staff to get the textbook and read it! Soon both parties were conversing not in requirements but in the almost intractable "real problems" that underlay them and what it might take to tackle them. In contrast to the six eagles story, this one had a profitable ending, winning a major development contract by responding in a language that both parties could understand — insights, questions, client satisfaction, and architecting.

Meanwhile, back on the academic scene

From an academic point of view, the subject has gained acceptance at MIT as well as USC. Most important, after 10 years it has continued to receive industry support of its students, now 60 per class in 1998. As a useful byproduct, the now-retired author–professor learned more from the reports by his students about more projects within 150 miles of Los Angeles than anyone might have predicted. The students generated, within the first 6 years, seven times as many insights as the professor had accumulated in 40 years in the field — an interesting commentary on just how limited each of us is in the opportunities to learn the lessons that insights can teach.

For more on this journey into the unknown, contact Dr. Elliott Axelband, Director of the Master of Science Program in Systems Architecture and Engineering, at *axelband@atlas.usc.edu* in the USC School of Engineering.

Summing up to this point

Part I showed that organizations were a form of complex systems and, hence, could be described in systems terms. Part II sketched how the world outside organizations and Part III how the world inside them could affect what the organizations could or could not do well. Much of this material is well known to excellent organizations, of course, though probably not in systems terms or priorities. Very little attention, for example, is paid to stock markets or natural catastrophes, though both have resulted in radical changes in industrial and international finance institutions. In system terms, they are treated as just two more outside forces to which the organizations' structure and its product line architectures must respond.

The purpose of the upcoming Part IV is, in Chapter 8, to describe the methodology of systems architecting and, in Chapter 9, what it can be expected to do for complex organizations facing change, illustrating both by architecting the start of a reorganization.

Notes

1. Profession: An occupation or vocation requiring training in the liberal arts or the sciences and advanced study in a specialized field. [WE 84]
2. The normative or pronouncement method is based on rules prescribed by a master of what is a "best" architecture. It is a sequel to the classical master–apprentice method of teaching. The rational method is based on deducing a good, even "optimum" design from applied science principles. It corresponds roughly to the systems engineering methodology. The heuristic method is based on insights generalized, or induced, from examples.

chapter eight

The context and use of systems insights and metaphors

Introduction

To understand organizational architecting means first understanding the basics of systems architecting, the subject of this chapter. The next chapter, Chapter 9, applies these basics to the architecting of organizations. The tie between the two is the product *line* of the company, be it hardware, software, consulting, or system integration. To produce that product line the organization must be appropriately if not contemporaneously designed, or architected, to do so efficiently and profitably.

Purposeful questions from inquiring minds

An appropriate way to describe the the architecting of *systems* is through a series of questions that students, managers, deans, and executives have asked over the past decade when trying to decide whether to study and incorporate architecting to assist in furthering their own interests and objectives, whatever they might be: broadening their career opportunities, more efficient managing of a development, establishing a curriculum, leading or managing an excellent organization. Whatever the purposes were underlying the questions, the answers did seem to help.

Most people who come forth with the most penetrating questions generally have heard of the subject through friends, staff, or associates acquainted with it and its possibilities. Most, like the reader at this stage, will have some understanding of its terminology and terms. The set of questions and answers that follow reflect many of the objectives just given. The answers in response are in two parts. The first part is brief, as were the questions; the second part, in brackets, is something like crib notes for the responder or "pointers" for a browser.

Q: What is systems architecting good for?

A: To help clients and architects answer architectural questions which applied sciences and mathematics cannot; namely, those for which there can be neither credible nor quantitative answers. [Typically, this situation arises because of insufficient information, of ill-structured problems, of too many variables, or of too many stakeholders with divergent values.]

Q: How about an example or two?

A: The easiest example is one basic to systems architecting, to decide whether the original statement of the problem is the right one. [The decision is a value judgment the client is responsible for making depending on capabilities, priorities, likely future needs, stated and unstated assumptions, and constraints. Architecting helps by asking appropriate questions and suggesting feasible options.]

The next one in the architecting a system — a product — is to decide on a combination of needs and feasible system elements that can yield a satisfactory solution; that is, an overall architecture. This process may well change the statement of the problem until a satisfactory combination emerges. [A matter of scoping, partitioning, aggregating, and assessing a satisfactory, not optimum, structure able to carry out the most important objectives of the product.]

Q: This sounds a lot like what some systems engineers call requirements analysis, a basic part of systems engineering. Is it the same?

A: Analysis cannot begin without at least one quantitatively analyzable concept in hand. Architecting, a joint effort with the client, is responsible for providing that concept and, during the acceptance phase, helps both client and builder come to agreement on waivers, if any. [See also ER 91 Preface. One of the most helpful short papers on this subject is LA 94 which, by past examples and in clear detail, shows the distinction in business information systems between their analysis and their architecting, and why each is necessary.]

Q: What are architectural tools and why are they different from, say, those of systems analysis?

A: Architectural tools are primarily qualitative *insights*; that is, generalizations abstracted from a number of problem-solving experiences in several different fields. Analysis tools are almost always quantitative and deduced from applied science and mathematics. Each is effective but for two different kinds of problems. [The most efficient application of architectural tools is before feasibility and desirability have been codetermined for at least one provisional solution. During implementation they help in managing the design and in providing sanity checks. During certification they assist in reducing contentions when "as built" and "as required" are not the same.]

Q: There have to be some issues around something new. What are they?

A: Briefly, the relationship between engineers and architects, credibility of the insights, and the role of the client.

Q: Well, then, how does an architect differ from an engineer?

A: An unanswerable and controversial question as stated. This question has been argued for at least 2000 years between (civil) architects and civil engineers, without defining a clear dividing line beween them. Less controversial is the question of how do the *processes* differ rather than the individuals. The answer to that question is easier — in the problems they solve and the tools they use to do so. [Organizationally speaking, the architect works *for*, or in the interests of, the client and *with* the builder. A typical engineer, on the other hand, works *for* the builder and, occasionally, *with* the client. For purposes of this book, a professional doing architecting is, at that moment, an architect. Similarly, a professional doing engineering is an engineer.[1]

Q: How do I tell whether architecting and its architects are doing well or not?

A: In the beginning of a project, observe whether desirability and feasibility are converging or wandering aimlessly. During implementation, see whether architecting is keeping the implementation on track or just chasing fires. During certification, judge whether the architect is providing useful input on waivers and helping or hindering in the resolving of disagreements between the client and the builder. [For further answers, see RE 91, Chapter 15, "Assessing architecting and architects."]

Q: What are the principal "technical" or day-to-day problems the architect works on?

A: Fit, balance, and compromise at system interfaces that affect the system as a whole. [As one of the famed classical architects, Christopher Alexander [AL 64], maintained,

The efficient architect looks for likely misfits
and designs the architecture so as to eliminate or minimize them.]

Q: In effect, doesn't architecting just "second guess" the builders and their work?

A: No, for three pragmatic reasons. First, no architect can be expected to know, in every field and application, more than the builder's experts. Second, the architect's difficult and time-consuming role is to focus on the system as a whole and on interfaces which largely determine system-level performance. And third, architecting, by professional consensus, does *not* engage in research, development, or project management unless by specific request of the client and with the concurrence of the builder. To do otherwise would put the architect in a position of conflict of interest and preclude earning the trust and confidence of the client and builder. [For further discussion, see RE 91, Chapter 1, "The systems architect."]

Q: Assuming that what you call "insights" are useful, how can they be *taught*? By case studies?

A: These questions are critical ones for educators and deserve more than a short answer. Indeed, the only credible answers must come from attempting to teach them. This has been done successfully at the University of California, but it did require the creation of a different teaching method. Its roots are in the time-tested methods of apprenticeship and "master classes," not of case studies. In the former, the order of presentation is lessons first, applications second. In the latter, the order is case studies first from which lessons emerge.

To answer the case studies question first, the wide diversity of the interests and experiences of the students meant that in each class of 20 — all of the students being graduate, full-time employed engineers and managers — 16 to 18 of them would quickly find that, in their own different contexts, domain-specific case studies would be largely irrelevant, if not obscure, and they would and did conspicuously lose interest. Further, in case studies the number of lessons learned from a single case study are very few and can be domain specific. Insights, on the other hand, are numerous and intended to be generalizable.

To answer the question of teachability: yes, it is possible and effective. The teaching technique is, in effect, the reverse of the case study method, or, if you prefer, a different starting point in the cycle of lesson, application, lesson, application. A list of generally applicable heuristics — the systems architecting term for insights — is presented in the first class period. The students were asked to accept them, at least provisionally, and to write reports on what happened when they were applied to each student's own project. Frankly, the results were astonishing not only to the students, but to their associates back at their projects, and not least to this author–professor. Admittedly, grading 20 completely different, 15- to 60-page reports in a week was not easy. Some students managed to use dozens of heuristics, most just a handful. Their managers were enthusiastic — something directly useful within the first month and a half. Clearly, something new was being added to what the students already knew in their own environment of which the professor could know very little.

Following this first report, from which the professor learned almost as much as the students, they were then asked to return again to their projects to see if they could find *additional* heuristics discovered in the course of the project. Having given the students about 100 insights in the first meeting, this professor was not anticipating very many more. But again, a humbling surprise to the professor and a welcomed byproduct. The students discovered many more — almost 700 over the next 6 years. Understandably, many were variations of the original insights, reexpressed in the languages of different applications. But even more were quite new to the professor, and equally "obvious!"

Teaching systems architecting, as measured by knowledge transfer, works. Better yet, it works in both directions.

The context of architectural insights: a quick review

The purpose of this section is to define the context within which architectural insights should be expected to work.

As noted earlier, the "common sense" of architecting is actually *contextual* sense; in other words, sense in a particular context. Whatever the context (field, discipline, or project), the principal focus of system insights is on the interrelationships among the constituents. Fix these and most problems are solvable if not solved.

Interface architecting in systems

Considering that interrelationships usually have more effect on high-level system capabilities than any other factor, they receive surprisingly little design attention from subsystem builders until systems integration, long after the systems have been sudivided into the subsystems. Builders' attentions quite naturally are focused closer to their own central issues. Even the best builders tend to forget interfaces and intereactions until systems test when, for the first time, "system troubles" cause delays and costly fixes.

For example, it has long been assumed by subsystem managers that almost any two subsystems can be connected to each other without serious difficulty. After all, fixing the occasional mis-fit in the past required only a few software programmers and a redesigned interface. Many managers seem to believe that as long as electrical plugs and connectors fit each other, all should be well. No longer. With the advent of distributed, domain-specific computing in each subsystem, it is not only possible but *probable* that the software architectures[2] of each subsystem will be mutually incompatible. They not only won't fit, they cannot be made to work.

As a practical illustration, if two subsystems are connected together, each with a central control form of software architecture, incompatibility is virtually guaranteed as the two contend for control of elements of the other as soon as the system attempts to operate.

In fact, Christina Gacek [GAC 98, see also AB 97] has shown that literally dozens of such mismatches, some considerably more subtle than the central control example above, will occur between different software architectures. Needless to say, software incompatibilities can have serious programmatic consequences, especially in software-intensive management and control networks. In one case, the development of a major space rocket required two different restarts from nearly completed software designs. The first restart came after the prime contractor failed to give the software contractor all the relevant requirements. The prime contractor then took over the software design but didn't have the core capabilities to accomplish the job. The government then ordered the prime contractor to re-subcontract which required the second restart. The total cost was about one billion dollars and 2 years in final delivery of the rocket stage itself.

This example is not unique by any means. The same collapse and restart has occurred in systems for the Internal Revenue Service, Social Secuity, police communication, state audit systems, company-wide finance and design systems, and many other applications. Characteristically, clients had been attempting to bring disparate hardware systems together using software "glue" when, in actuality, the glue was the crux of the overall system design problem. This problem has now reached the point where software is specifying what the hardware must be, instead of the reverse. The present design system of hardware-first, software-second has forced software costs to exceed the costs of the hardware. Balance will soon, if it doesn't already, mandate software priority in many system designs. Fortunately, software systems architects are some of the most vigorous and innovative systems architects in the business.

What does that have to do with organizations? We may find that trying to bring together two historically separate entities, from groups to enterprises, will expose incompatibilies that are very difficult to resolve.

The use of architectural insights

At long last, after so many chapters and issues, the first payoff — an organized set of tools for managers and professionals to treat practical problems that can't be quantified, measured, or replicated; a storehouse of architectural insights with metaphors to help understand them.

Definition

Webster II [WE 84] defines "insight" as the capacity to discern the true nature of a situation, literally a look (deep) into a situation or structure. As used here, an insight is a phrase or graph that gets to the essence or nub of the matter. When first encountered it gives a new perspective or slant to one's world, one which can abruptly alter that world — an "Ah hah!" or "Eureka!"

Insights are the foundation of what architects and other professionals call "common sense, instinct, and creativity" in describing how they do what they seem to do so easily. The best of the professionals are continually adding, modifying, and reorganizing the insights that make up their "tool kit," as some call it.

It is common, and probably wrong, to equate insights with born talent. Skillful use of insights can make for fine architects. Adding talent to any skill, including achitecting, makes for great ones, as in any profession. Architects, indeed, are much like other professionals — doctors, lawyers, statesmen, and artists particularly — all of whom live in ill-structured, nonquantitative worlds and require a high level of perceptive insight to be successful.

C. W. Sooter [SO 93] may or may not be quantitatively accurate but he is certainly insightful in hypothesizing that:

> *An insight is worth a thousand analyses.*

Insights come in different forms including descriptive heuristics that describe situations, prescriptive heuristics that suggest how to handle them, and metaphors that attempt to make arcane subjects more understandable. Also shown to be useful are insightful reference architectures and project reports that provide deep insights into the projects which first "discovered" them.

Heuristics

Definition and form

Heuristics as used here are guidelines — generalized lessons learned from experience, simply expressed. In sheer numbers, they are by far the largest number of tools in the architectural "artists" tool kit.

The readers of this book by this point will have been exposed to more than five dozen of them, inset and italicized. And there are dozens more to come. Several things can be said of them immediately. They can quickly summarize whole paragraphs, sections, and management papers (and sometimes whole books). They are best appreciated in context rather than, say, in a long list. Most of those given in this book are intended to be used throughout the organizational architecting; that is, conceptualizing, partitioning, evaluating, organizing, merging, and so on. As earlier examples have shown, there is a great deal of "carry-over" of insights from one application or field to another, particularly from engineering into managing and vice versa.

Credibility

The subject of the introduction to Part IV, and the last question in the Q&A section that begins this chapter, concerns the teachability and credibility of insights. Why should the reader treat them as more than the off-hand pronouncements of a single author?

Well, first they represent the experience of dozens if not hundreds of individuals in many fields. The author has attempted to certify and then report them and, if they pass the criteria for credibility, to supply some well-known ones in fields in which he is knowledgeable. The most recent source has been the hundreds of reports by systems-experienced graduate students describing their experiences in and "lessons learned" from their many diverse projects.

Second, a surprising number of people on encountering a valid insight see it as "obvious." They will quickly pick it up and use it in all kinds of situations. The insight is credible to them because it fits in their frame of reference, in their experience, and for their problems.

And third, knowing about them — their nature, utility, and credibility — can help keep the user from being overwhelmed by columns of numbers in an accountant's or analyst's report. There is, indeed, nothing quite like a simple sanity check or an insightful question to find the source of likely flaws in a nonintuitive conclusion. The Five Why's at first may seem trivial but they are extremely powerful as working techniques.

There is one more reason for believing them. Experienced students use and enjoy them. Two of their favorites to date are the famous:

> *Don't assume that the original statement of the problem*
> *is necessarily the best or even the right one.*

and the powerful architecting and design "rule,"

> *Simplify. Simplify. Simplify.*

which many regard as one of the most useful responses to Murphy's Law:

> *If it can fail, it will.*

The real objective of teaching architecting

By far the most frequent comment by students at the end of the systems architecting course is, "I think differently now," to which some add, "and I cannot go back." This quite involuntary, unsolicited comment suggests that the best statement of the objective of teaching architecting is to help students think insightfully; that is, to help them search out, retain, and use insights as guidelines and touchstones for thought.

The definitive work on the recognition, selection, and application of heuristic systems architecting is that of Kenneth L. Cureton [CU 91]. Remarkably, its suggestions are expressed as heuristics about heuristics, or what Cureton calls "metaheuristics." [CU 91]

Sources

It is clearly impractical to report on all possible sources of organizational insights. The principal ones have been from peers, associates, and friends; from project and program reports; professional papers in peer-reviewed journals;[3] and, of course, from reports from graduate students in systems architecting. Probably to avoid the appearance of speculation beyond what has been accomplished, few writers highlight in their articles what may be more widely applicable insights. In fact, generalizable insights don't often appear in either the abstract or conclusion but remain buried in mid-text. The exception is the USC graduate student reports, all of which were required to list in a special section what the authors thought were generalizable insights.

To illustrate, one of the best student sources for organizational insights, both in number and quality, is Meg Renton's *Organizations as Systems and Architectural Heuristics* [REN 95][4] which, against a background of planning the future for McDonnell Douglas Aerospace, gives a set of new lessons she learned from that particular experience.

1. It is less painful to anticipate than react.
2. Anticipation takes resources and commitment. Reaction loses both.
3. The reverse of diagnostic techniques are good architectures.
4. Structure is but one element of the architecture.
5. There is no single best architecture. It depends upon the environment.
6. You must know where you've been to know where you're going.
7. Be aware that architects slant architectures toward their specialties.
8. It is not enough to have visibility into the misfits...you must first understandand then do something about them.
9. A vision is an imaginary architecture...no better, no worse than the rest of the models.
10. The optimum number of architectural elements is the amount that leads to distinct action, not (just) general planning.
11. 100% fit is not only highly effective, but highly improbable.
12. Poor aggregation results in grey boundaries but red performance.
13. Unless time stands still, no architecture can stand the test of time.

Of these, the most widely useful insights would seem to be: 1, 6, 9, 12, 13, and possibly 3. The wider significance of No. 3 would be more evident if rewritten in short, but perhaps too intellectual a form as "the reverse of *a posteriori* assessments is *a priori* architecting." At least half of Renton's list have been, or can be, found in other organizational settings, indicating the universality essential to a valid insight. Readers are invited to make their own selection if for no other purpose than to understand how insights can be chosen for personal use.

Some insights have a long and acclaimed lifetime. The foregoing *Simplify. Simplify. Simplify* is one such and may well date back to the time of the Egyptians. In the aircraft industry the same idea is expressed in a question and answer form. The question: what is the most reliable, least costly, and best scheduled part on an airplane? Long pause... The answer: the part that isn't there, because it isn't needed. [Hillaker 1989] Another prescriptive (design) heuristic from Hillaker's F-16 experience is

Every part must earn its way on to the airplane.

As was shown in Chapter 2, the widely known but seldom implemented Murphy's Law and its corollary (*If it can fail, it will — so fix it first!*) summarize the difference between the Japanese and American automotive industries. They also encapsulate the principal management strategies for the remarkable performance of the Apollo lunar program, the long lifetimes and profitability of communications satellite networks, and the astonishing photographs and radar presentations of the moon, the planets, and the Earth. These insights are in strong agreement with the zero defects strategy of those programs.

Metaphors

> "The essence of metaphor is understanding and
> experiencing one kind of thing in terms of another.
> Its strengths are the inherent similarities of both." [VE 93]

Definition

A metaphor consists of a matched pair of two objects or systems that are generally similar in behavior and/or purpose, but are found in two quite different disciplines, fields, professions, or projects. The purpose of a metaphor is to educate nonspecialist stakeholders by describing the characteristics of the less familiar in the terms of the more familiar. By similarity or analogy, key concepts are brought across from a more arcane subject to an everyday one in an effort to make the arcane more understandable, believable, graphic, and acceptable.

At least a dozen metaphors have been used in this text so far from eagles to desktops, from boats to water flow, and from braided ropes to level playing fields. Their consistent purpose was to educate newcomers to the field.

Can metaphors also be used to solve architectural problems directly? As a matter of fact, a metaphor was indeed used a few years ago to deal with a serious problem presented by what had become overwhelming numbers of insights. Their rapidly increasing number made it clear that they could not be combined into "meta-insights" from which the simpler insights could be derived. Despite the plea of a famous but overburdened chief executive officer, there were not going to be useful "7 ± 2" master insights for organizational success from which all other insights could be derived.

A simple metaphor, the hardware store, solved the problem. Imagine going into a hardware store with a problem, say in plumbing, electrical appliances, carpentry, painting, or lawn sprinklers. You enter the store and see that the manager has organized his wares in aisles marked plumbing, electrical, and so on. You go to the right aisle and find hundreds of tools, fixtures, and supplies. You wander down the aisle automatically ignoring those products you don't need in a search for something that might help, say a hammer, nails, and a saw. Did you know that there are 150 different kinds of hammers and dozens of kinds of saws and nails? So, you pick out a hammer from the hammer shelf that looks useful, a saw with the right kind of teeth and length, and some nails from the nail bins that are of the right length and type, and you have all you need. A small selection — maybe even 7 ± 2 — of a few items has solved your problem, so you buy them and leave. You do *not* have to know all the products in the store, much less have to buy them all. The situation is much the same for selecting insights, requiring only that they be organized in aisles, shelves, and bins for easy browsing, selecting, and taking home to your problem. Such is the "hardware store" of Appendix A, its "aisles" organized by user interests or objectives.

In effect, the hardware store metaphor made it possible to respond to an overburdened CEO by saying, "Tell us the kind of problem you have and your architect can bring you the handful of insights of most value in resolving it." This response should be no surprise. Engineers do the same thing all the time. There are no 7 ± 2 *wunderbar* principles there, either.

Form

Creating useful metaphors clearly requires a different kind of insight from that of formulating heuristics. Metaphors require an insight so deep into both the proposed system and its metaphor that the behavioral characteristics of the two are almost identical in at least two levels of detail. When that is possible, then the metaphor can be reliably used by the stakeholders to anticipate likely behavior of the proposed system. The objective, of course, is to make sure that the promises and constraints implied in the metaphor are not only similar and desirable but are comparably *feasible* for what is proposed.

For example, the desktop metaphor for personal computers was so obvious to even first-time users that few bothered to read more than a few pages of the operating manual before plunging in. It not only looked like a desktop, it worked like one — only better and smarter. Few knew, or needed to, what microchip did what, or even that microchips were employed.

At the same time, and this was sometimes missed by the manufacturers, for the computer to do *more* than the desktop metaphor could be disturbing if not frustrating for the user. In a sense, the Macintosh iMAC is a pullback from what a computer *could* do to what it *must* do on a desktop. A metaphor can also do damage if what had been characteristic of a system is radically changed. The "Ma Bell" metaphor that characterized the ever-helpful, always-ready, user-friendly Bell Telephone System rapidly degenerated when the Bell System was converted by government order into a profit-comes-first gaggle of suppliers which in turn converted a pay-as-you-use architecture into a pay-for-access billing architecture. It also became a planning disaster to the Defense Department which could no longer turn to a single partner for defense communications. "Ma" was missed and those whom she had served best have missed her the most.

The lesson for excellent companies contemplating radical change:

Watch your image!

It, too, can affect what you may be able to do or not do.

Metaphors as commentary

Some of the most famous metaphors appear as cartoons — images intended as political, social, or managerial commentary. Satire, bitterness, sorrow, condemnation, and celebration can all appear in a single metaphorical cartoon. One of the finest political cartoonists, Paul Conrad of the daily *Los Angeles Times* for more than 30 years, is reported to have claimed, when asked about it, that his endlessly creative cartoons were generated by a very simple technique. He would look at the previous day's front page, pick two subjects from it almost at random, find the underlying similarities or coincidences, use the resulting metaphor, and start drawing. By whatever means Conrad generated his metaphor of the day, they have been extraordinarily powerful in molding public opinion on major national issues. Without question the "fuzzy familiarity" of his metaphorical cartoons was much more effective than many pages of commentary, analysis, and opinion.

Even closer to the issues of organizational architecting are the cartoonist-engineer Scott Adams and his "Dilbert" series on management and its foibles. Management would be well advised to pay attention, even if affronted and disturbed by Adams' deliberate exaggerations, to what metaphors can do to the image and future of an excellent organization, the merger of two metaphorically different organizations, or the morale of a company's employees. And, hence, to what an excellent company can or cannot do well.

Sources

Metaphors are the stock in trade of writers, commentators, visionaries, and multifaceted architects. Indeed, many dictionaries give "figures of speech" as the most common usage of the word "metaphor." However, as noted earlier, to make them useful and credible in organizational architecting these figures of speech, logos, and mottos must be closely tied to the behavior and purpose of the organization. That need implies a deep knowledge not only of the present and desired future of what the company can or cannot be, but also a real sensitivity to the perspectives of its stakeholders. The conclusion? It would seem that the most helpful sources would have to be close to, if not within, the organizational architecting team. The metaphors, if any and when needed, would be conceived as part of the evolution of the company architecture, policies, and procedures.

Extensions of the field

The field of systems architecting has been extended well beyond heuristics and metaphors, as shown in the introduction to Part IV. Of the extensions, two of most applicable to architecting organizations, the subject of the next chapter, arguably are Cureton's Metaheuristcs and the efforts of Madachy, Frericks, and Maier to bridge the gap between the art and the science of systems architecting. Generally speaking, their results indicate that connections across the gap can be made as long as the complexity, or number of interrelations, is not too great, perhaps up to 50.

A disturbing development for organizational architecting is the inherent incompatibility of software architectures as demonstrated by Abd Allah and Christina Gacek. By implication, the elements of information-intensive organizations may prove difficult to interconnect in a way that yields emergent or synergistic capabilities. They may not be able to talk together constructively.

Summary

This chapter begins with the questions often asked by people with an expressed interest in systems architecting, from what does it do to how it can be taught. It shows that the skilled use of insights is the basis of what architects call common, or contextual, sense. Heuristics, "the intelligent response to recurring themes" [OL 92], and metaphors, easy-to-understand analogies for describing the characteristics of more difficult architectural concepts, are shown to provide tools for dealing with the not-science, not-mathematics problems treated by systems architecting.

Notes

1. For further answers, see RE 91 Chapter 1, "The Systems Architect" and [LA 94].
2. Examples of inherently different software architectures or styles: pipe and filter; data abstraction and object-oriented; event-based, implicit invocation; interpreters; process control; and blackboards. [From SH 96]
3. For example, a single paper, mostly on heuristics for spatial search [ST 94], lists 943 *more* sources.
4. Used here with her permission.

chapter nine

Architecting a reorganization and forming its architecting team

Reorganization cannot be avoided; it is a natural part of growth.

Introduction

The purpose of this chapter is to illustrate how architecting works using reorganization as a practical example. Before walking through the re-architecting process, two questions should be answered. Exactly what is being re-architected? And, who is the client and who is the architect?

What is being re-architected?

In effect, two things have to be re-architected at the same time, the organization and the product line it produces, because ideally:

> *Th architectures of the product, its construction process,*
> and the organization that manages them
> *should "fit" each other.*

This insight is a relatively new one. Internally consistent systems architectures have been around for at least 50 years. Well managed processes, usually based on the hardware "waterfall" began the industrial revolution. The integration of products and process designs, though much more recent, have considerably improved corporate parameters of quality, time-to-market, and profit. Examples include planned product improvement, lean production, and concurrent engineering. It should not be too long before the architecture of the organization, itself, joins the crowd.

After all, there is no reason to assume that the hierarchy is the only and best architecture for the creation and building of software-intensive products, much less of knowledge systems. There is no reason to assume that all businesses should be run the same way just because one group of stakeholders, the stockholders, happen to share similar interests across a number of product types. Indeed, in a knowledge-producing business, there is no reason to have any material assets at all; everything can be leased and the biggest asset of all, its people, can walk off and form their own company any time they wish.

A corrolary of the preceding insight is that reorganizing the organization implies re-architecting of the product and vice versa. This subject is sufficiently important on its own that it will be revisited at the close of the chapter.

Who is the architect and who is the client in a reorganization?

Assuming that re-architecting of product, process, and organization are considered together, the next questions, and they are nontrivial, are who is the client and who is the architect of this process?

Although there are exceptions, the clients are those that make value judgments and the architects are those that provide answers to feasibility questions. In small organizations, both functions may be performed by a single individual, but as organizational size and complexity become greater, the focus of value judgments increasingly becomes top management. It is the only location in the organization with the breadth of view, both internal and external, to do that job. Perhaps needless to say, making value judgments — what is good or bad — is neither a staff job nor one for popular vote. Consensus helps, of course, but not votes. The stakeholders are too diverse, even among stockholders. In some organizations, the CEO alone takes on this responsibility. In others it is more collegial among the top managers. But it is still at the top and will be assumed so here.

The question of the architect, the feasibility advisor, is more complex. Conceivably it could also be the CEO alone, a leader with a vision and the knowledge of how to achieve it in the interest of all the stakeholders. There have been such and no doubt always will. If that is assumed to be the case then, in what follows, the CEO–architect will be both posing and answering the questions in an architecting process that requires both client and architect to participate.

More generally, the architect will be an architecting team headed by a chief architect, a team chartered to do the job as objectively as possible and with as little personal or organizational conflict of interest as possible. If this is the case, then top management is the client and the chief architect is a professional on the staff or from an independent partnership.

In any case, the end objective is to structure the organization to add value to the end product to the benefit of all stakeholders.

Architecting a reorganization

Following a technique developed by the author at the University of Southen California for the purpose, readers, acting as clients, will make all judgments of what is or is not worthwhile as the architecting process develops. For example, in the very beginning readers will have to judge whether a reorganization should even be considered in an already excellent organization.

The author will act as an architect, asking questions until the reader, the client, can decide on how and when to act, if at all. The architect will follow the established ethics and principles of architecture — not to recommend nor to respond to the question, "I'm confused. What would you do if you were in my place?" — except with further questions. The architect's questions are inset and indicated by a "Q." As in systems architecting, there are no right answers. In this case, there are *no* answers. The author assumes only that the clients' own answers lead to the next question. The architect in this example acts only as an objective mirror, showing the viewer what others see so that the viewer can decide on what actions to take.

From time to time the author will interject an architect's perspective on the status of the process as it moves along. Following the reorganizing process, the chapter addresses the closely related issues of the creation, composition, and use of an architecting team, and the close relationship between architecting an organization and architecting a product-line.

It is very important in what follows that the reader place the process in a specific, preferably personal, organizational context because the text deliberately does not provide one. This chapter, as will be seen, is *not* a "how-to-reorganize" pamphlet illustrated with a few case studies. The author freely admits that his own experiences in reorganizations have been used to frame many of the questions. But many were also the result of a frank and insightful description of the re-architecting of the Xerox Corporation's organization and along with a major expansion of its product lines in the early 1990s.[1]

So, as we begin, please now choose your own context. There are no suggested or implied responses that you might have made — it is your task to make them, remember them or note them in the margin, and keep going until you know the answers to your own questions.

Depending on your choice of organizational context, you may find at some point that no further questions are needed. The mutual architecting process and you, the client, will have provided the answers you need. *The architect as a mirror will not have provided the answers.* The rest of the process given here is then for clients of more ill-structured or complex sets of problems, say, those in which operating issues or "turf problems" as they are called, may impose further constraints on the reorganization plan.

The process begins with an introductory set of questions that allows both the architect and the client to assess each other. There is an arguable reasoning that says that if the client and the architect do not come to an easy and mutual understanding, even liking, of each other, either the client should fire the architect and find a more congenial one or the architect should arrange a tactful withdrawal. Something like finding a good doctor, no? To those who might object to what may seem to be a heartless decision, it is by no means uncommon. If the client prefers a product-oriented, profit-making style of management and the architect is much more familiar with a collegial, service-oriented one, mismatches and misunderstandings will plague the process throughout. Better for both to shake hands and part company as friends. Yes, this author has had experience down that path as well.

The first step: acquaintance with each other and the client's presumed problem

Q: Thank you for asking that we meet. You know my architecting background. You told me that you were thinking about a reorganization and thought I might be someone to talk to. How can I help? First, to help me, can you tell me something about your business? What's its history, particularly its beginnings? Is today's organization more or less the same now as then? What kind of a customer base do you have? How many different products? What is the average experience of your managers? Yourself? In this business, alone? Have you enjoyed being here? What have you enjoyed the most? Everyone has their aches and pains in a growing business. What were yours, if any? What kind of a management style do you and your immediate associates, above and below, have or prefer? Collegial? Strict lean and mean line authority or egalitarian and relaxed? Who are your major competitors? Your customers? What do you regard as your major successes as the excellent company you've become over the years?

Q: Sounds great to me. Why are you thinking about reorganizing at all when all is going so well? Is the competition reorganizing? Are there standing problems that need to be fixed? Did key people leave or retire? Do I sense that maybe some will or should? Or is it just a good opportunity to shake things up? I have a friend in business that believes every company should reorganize every 3 years just to "break all those lesions" as he puts it. Would you agree with him or prefer not to change horses in main stream without very good reason?

Author's interjection for the reader: Note the use of graphic metaphors (lesions and horses) to elicit a context-free response. The purpose of the last set of questions is for the architect to learn just how personally involved the client has become, or intends to become, in the reorganizing process and its rationale. (Founders tend to become intensely involved which can mean the imposition of idiosyncratic constraints that must be accommodated or the process will collapse.)

It is critical that the architect work with the individual or group that makes the value judgments, definitely not with a decisionless staff assistant assigned to gather ideas. Otherwise, further discussion is likely to be a waste of everyone's time. If the architecting process is on track, enough trust should have been built up between the value-judging client and the architect that both can speak freely and honestly without the need for intermediate interpreters of both client and architect ideas.

Typically, both should have an understanding that relative importances of the principal stakeholders are, more or less in order, the customers, the company as a viable entity, its people, its reputation, and so on. Missing a principal stakeholder at this stage can create serious acceptance problems later on. "I wasn't consulted, so I cannot very well agree" can stall any process.

At the same time, the several purposes of the reorganization — their relative priorties, the criteria for the success of each, and their detailed implementation — will (or should) still be provisional, hazy, and subject to change without rancor. It is simply not possible for any client, no matter how intelligent and experienced, to answer questions of purpose, priority, and intent without having some idea of what the end result might be like. Ask a typical client what is more important — performance, cost, or schedule — and the only possible answer at this stage is "yes." The as-built implementation inherently includes within its design the *actual* relative importances. Reflecting the reality of what is feasible, they almost always are different to some degree from the original hopes. As a popular heuristic states:

> *As for performance, cost, or schedule, pick two*
> *and reality will determine the third.*

Thus, at this stage, it is wiser not to lock in any one of them except in the broadest terms.

The second step: determining what is both desirable and feasible

Q: You have said that the customer and your survival must come first. That usually implies a possible change in product line. In other words, is adding value to the customer a driver in this case? What value? Is that an easy sell or something that may take some educating and selling? Are there preferred customers — the 20% that buy 80% of your products, so to speak? How would a reorganization affect them?

The client at this stage has probably not thought too much about such key questions. Now, thinking about them has resulted in discarding whole sets of possibilities — those that would not add value as perceived by the customers (and still other stakeholders, including the stock market come to think of it). This realization can be refined further to acknowledge different values of different stakeholders, e.g., company profits, perceived ethics, so-called "good will," company loyalty, and so on. More than one opportunity has been squandered by transferring the favorite tech rep or sales manager of a key customer to the account of the customer's competition. The remaining questions can finally be aimed at the proposed reorganization(s). It is only now the time for the architect to make some presumably feasible suggestions (*not* recommendations!), fully expecting that some will be rejected outright. There is an insight here:

> *"I hate it!" is direction.*

The architect may never find out exactly why, but:

> *Don't buck it or even try to!*
> *Learn from it. If the client reverses course,*
> *don't make a big deal of it.*

Everything at this stage is provisional.

Q: With this growing (or decreasing) number of product lines and individual models, you might be heading to a matrix (or project) structure. Have you thought about this? Have you thought about which functions and projects might go where in the interests of increased value to them? Have you or any of your associates proposed any particular structure that at least deserves consideration at this stage?

Q: I can see you have. The advantages of each seem to match experience various organizations have had with them. It is much harder to dig out their disadvantages, though. Few organizations like to talk about them and even the best of the media don't always get the situation quite right. Can you have a discussion with your associates and develop corresponding lists of the disadvantages? Try hard to be as critical of what might look like the good ones as of the bad ones. Be sure to include your own present one. Be sure you include what your customers might think. Don't prejudge any of them if you can help it. There is still a long way to go. Because:

> *The choice between the possibilities may well depend upon*
> *which set of drawbacks you can handle best.*

and that is going to take a bit more time and thinking to sort out. Remember as you go that few companies fail because of their assets. Most fail because of management's inability to mitigate built-in drawbacks. And, to reiterate the theme of this book, it is often the drawbacks that directly or indirectly determine what an excellent organization can do.

Let's assume the reader, or client, has done the necessary homework and has concluded at least the first search for a possible solution *including* not reorganizing. Chances are, in the latter possibility, that a number of constructive suggestions will have arisen that indeed do increase value all around. If so, the real objective has been realized and the architecting process has been a success.

If, on the other hand, a specific kind of reorganization has been chosen, often considerably improved from its first version, then it is time for the next step which is *not* to make a big announcement and charge ahead.

The third step: resolving implementation issues

There are several very good reasons for not committing yet to the (provisionally) chosen reorganization:

- Odds are that the criteria for success have not been worked out.
- Odds are that the acceptance criteria have not been worked out, either.
- It is too soon for analysis. It is time for insights to surface. Refinement by analysis comes later.

and so, one of the most important insights in this book:

> *In introducing technological and social change,*
> how *you do it is often more important than* what *you do.*

The record shows that 75% of all attempts to introduce automation into the workplace have failed, not because the technology was deficient but because the introduction was mismanaged. [MA 88] Reorganizing is a minefield and it is best to clear out the mines, if possible, before marching on. Many a building has been architected that could not be built, not because it couldn't stand up but because there was no practical way to assemble it. As they (used to) say[2] in software architecture:

> *Be prepared to throw the first one away. You will, anyway.* [BR 75]

The intent of this insight here is to warn against being too closely tied to a particular option too early. The single question common to all these insights is "does the reorganization follow these guidelines and does it add value to the organization as a whole?" Other questions will depend upon prior client answers, such as the judgment that a matrix (or a project) configuration seems best, so far. For example, an interface is prone to more errors of omission and comission if it is far removed from where value judgments are being made, and that depends upon the structure of the now seriously proposed reorganization.

The fourth step: a careful look at key interfaces

The essence of reorganization is the aggregating and partitioning of its elements. Aggregating brings like, or closely related, things together. Partitioning, the other side of the coin, keeps dissimilar or distantly related things apart.

Organizational aggregation is the less difficult of the two. It is common in downsizing where it is almost unavoidable. It is the delight of empire builders and authoritarians. It is the natural inclination of advocates of a new concept who tend to bring every issue under their umbrella, an inclination that often sinks the concept under its own weight. Such is the case for commercial off-the-shelf (COTS) usage, a valuable idea within a limited context, a disaster for a complex system. The expression most used in looking back at an over-aggregation is "just one more passing fad." Though no longer a concern of the (civil) architecture profession, now thousands of years old, it is a present one for *systems* architecting. It is because of this concern that the principles of systems architecting are so strongly patterned after that of its ancestor — firm resistance to over-aggregation, strong ethical prohibitions against perceived conflicts of interest, and, admittedly, professional self-interest — the keys to the sustained health of architecture. This book is written in this tradition.

Partitioning is far more difficult, particularly in complex organizations because of a common (but erroneous) assumption that authority should match responsibility, a condition that is almost impossible to fulfill in any sizable human organization. But that assumption is the cause of considerable pressure being exerted by the affected parties for "clean, definitive, workable, and enforceable" interfaces between them. In any system or organization in which value is found primarily in multiple interfaces, those demands or policies are just plain unrealistic. Cooperation is what is required, but measuring it in overall system parameters is at best a subjective exercise. Like success, itself, cooperation is in the eyes of the beholder.

The difficulty of determining and implementing the interfaces is further compounded by demands or policies calling for personnel evaluations based on measurable criteria. If the interfaces are really hazy zones of mutually conflicting interests, a misjudgment there can paralyze an organization. For example, discrimination, preferences, affirmative action, and tort liability claims all stand astride the interface between a human resources department and a legal office. If combining them would only generate still more difficulties, where should the interface between them be and how should it be implemented? Some of the most useful insights in handling such partitioning problems are

> *Relationships among the elements are what give*
> *the organization its added value.*
> *The greatest leverage, risks, and dangers are at the interfaces.*

This pair is a reminder of the importance of interrelationships and the need for care in establishing them. These can then be followed by generally accepted ground rules:

> *Do not slice through regions where high rates of information*
> *exchange are required. In partitioning, choose the elements so that*
> *they are as independent as possible; that is, elements with low*
> *external coupling and high internal cohesion.* [Gold, Jeff 91]

Most managers of organizational units will welcome the above on sight. And they will be glad for the autonomy justified by:

> *Organize the sections, etc. to make their performance*
> *as insensitive as possible to unknown or uncontrollable*
> *external influences as practical.* (Huch, David 91)

And, finally, a comparable insight that explains the architect's needs and purpose to managers and stakeholders alike:

> *Try to determine which interfaces are most error-prone.*
> *The efficient manager, using contextual sense, continually looks*
> *for the likely misfits and redesigns the organizational structure*
> *as to eliminate or minimize them.*

A note of caution: as with all insights, these require judgment and balance in their use. For example, the foregoing ones could result in so much autonomy of the units that external direction, much less change or cooperative action, could be seriously impeded. The simplest antidote for over-use, once again, is the systems concept of decisions based upon the effect on the system — the product and the organization — as a whole.

Rather than carry on the question format of the earlier steps, by now understood in principle by the reader, the reader is now encouraged to role play the architect and to take each of the insights and formulate two to four penetrating questions for those who must respond to them if the reorganization is to have a chance. This switch in roles has proven to be a favorite of students in understanding the complex human relationship that must exist between clients and architects.

The fifth step: assuring effective operations well ahead of time

Operations differs from earlier phases of reorganizing in that it invariably involves the outside world. To this point, with the possible exception of compiling the list of stakeholders, reorganizing has been an internal affair. Now it is largely external — the media, the government, the unions, the new generation of customers all of whose perceptions can vary from enthusiastic to brutal. It is too late by the time the effects of a reorganization are felt in the marketplace to do much beside damage control. The external world must be incorporated early, certainly before the end of Step Two. Success should have been defined; acceptance criteria (like maturity indices, market share objectives, initial sales rates, and the like) established; and options built in for quick response to market reaction to the company's action *before* the reorganization is announced, much less implemented.

As the reader no doubt has noticed, the architect's role is greatest in the early steps and diminished in this last one, except, of course, for helping assure that the external world is indeed incorporated early. Waiting too long may mean starting over.

This chapter now returns the reader back to the first step to include there, much as a skilled architect would, other closely related factors that affect the success of any reorganization: forming the architecting team and coordination with changes in product-line architectures.

Establishing and working with an architecting team

The architecting team's responsibility to the client

The architect's primary responsibility in a potential reorganization is to help the client come to a desirable and feasible answer to the questions of whether and, if so, how to reorganize. To the extent that quantitative analysis is required, the architect should formulate the client's answer in an analyzable form — an architecture. Most clients at this point will need some friendly, trusted, technically competent, and economically disinterested assistance not only at the value judgment level but also at the implementation level, as more detailed questions arise.

One of the better sources is an architecting team who can help with the leg work, clarify the statement of the questions and calm down anxious stakeholders, particularly employees and key customers, and so on. The formation of an appropriate architecting team with a well-understood charter of do's and don'ts can considerably improve the chances of success. For the purpose of this case, it will be assumed that the team is being assembled from within the company with, perhaps, some facilitating assistance by consultants familiar with the architecting process.

The client's responsibility to the architecting team

Without question it takes courage by the top management to form an architecting team. The team no doubt will come to management with findings that the management would rather not hear of or discuss immediately with the team. The team undoubtedly will take more management time and effort than anticipated. Some parts of the company may demand the disbandonment of the team or at least a prohibition of it having any access to their part of the company, a demand that must be refused or the reorganization may fail by omission, which may be precisely what the demanders intended. This is not only an idle possibility, it is a probabililty.

Such indeed was the case in the late 1940s when James Forrestal, then Secretary of the Navy, built such barriers against the newly organized Department of Defense that when he was subsequently appointed Secretary of Defense, he could not bring the Navy into the DoD scheme of things. As the expression went for some decades: the Navy never joined the Department of Defense. It never recognized tactical vs. strategic partitioning, defense vs. offense weapons distinctions, nor the DoD's control of the budget. No wonder

Admiral Rickover could bypass it completely and deal with the Congress as it had for the previous 150 years. As it turned out, both the Navy and the Department of Defense suffered as a consequence.

What does this have to do with forming an architecting team? The lesson for potentially affected parts of an organization is to participate, to provide well-qualified members if and when requested, and to be where the action is. The lesson for the team is to demand and receive open access to all parts of the organization as needed for the enterprise. The lesson for the top management is to form a team, charter it well, support it conspicuously in public, and be willing to provide it with implied authority when needed.

This lesson is crucial because the architecting will face formidable odds. According to research by Donnellon and Margolis [DO 94 3-14], any group in a fairly large organization will face at least five core dilemmas: (1) an inhospitable external environment, (2) a culturally different and possibly revolutionary structure, (3) a mismatch with the organization's orientation, (4) a post-team assimilation into the main stream, and (5) support dependent on results in the normally reluctant organization it is supposed to serve. Direct analogies can be found in military service in which officers are seconded to headquarters for planning activities, a duty almost uniformly disliked by the officers, and then have to return to their services with a time-out (usually negative) in their careers.

Although Donnellon and Margolis have suggested maxims to counter these dilemmas, the most important single lesson for the client is that forming and working in an architecting group is a delicate and difficult job. Lacking strong external support by the client, not only will the results be limited, the needed sources of expertise will shun the tasks. In the reorganization example given earlier, the architects can ask any questions they like but the necessary responses will be reluctant if not destructive.

In particular, the client should be prepared, prior to the establishment of the team, with at least a brief statement of the stated objectives of the enterprise, the present structure of the enterprise, its product lines and products, the assumptions that led to this study of reorganizing parts of it, and the presumed scope and lifetime of any likely reorganization. As the reader can understand, such a statement answers the first questions the architecting team is bound to ask. Both client and team should understand the provisional nature of the last two factors; that is

> *Neither client nor architect should assume that the original*
> *statement of the problem is necessarily the best,*
> *or even the right one.* [RE 91 54]

The client, prior to the announcement of the formation of the team, should have prepared a statement of the chartered responsibilities of the team and cleared it with the managers most affected for possible modification. The charter should contain:

- The purpose of the study: working with the organization under the supervision of the top management, study the desirability and practicability of a reorganization in the interests of (contingency planning, adding to the value delivered to the customer and enterprise, etc.).
- The intended scope of the study.
- The access granted the team to all (or selected) sectors of the enterprise.
- Deliverables (an analyzable proposal for further study, critical times, study duration, etc.).
- Team composition (by name and/or expertise and background).

Such a charter would help diminish rumors, calm management concerns, assist and protect the architecting team, its members and their future careers, and provide a channel to a supervising authority for official information and for expressing concerns, if any.

Understandably, this portrayal of the establishment and chartering of an architecting team is less than a roaring endorsement for the reader going right out and doing so. Reorganizing is traumatic for any organization and particularly for an excellent one doing well. It and its members very likely will take whatever protective measures they can to block or prevent it. If, and only if, its top management is trusted should it even be attempted. That mantel of trust is the team's best protection and justification.

This much does appear to be true:

- In times of rapid global and technological change, being prepared is better than standing still.
- If product line change is probable, reorganizing to help it succeed may make sense.
- It is certainly better to study the possibility before the storm, using the best available talent, than to be caught in a deluge out in the open. (Any metaphor in an emergency!)

The composition of an architecting team

Whether formed as a permanent adjunct of a product line subsidiary or as a temporary transition team, an architecting team must be composed with care or it will fail in its purpose.

As a first principle, architecting teams are, and experience confirms, relatively small, a handful at most.

Too many cooks spoil the broth — and get in the workers' way.

In architecting, bigger is not better. The need is for a very good architecture that does what is needed most, not an optimum that provides the holy grail for everyone. The small size, however, is more sensitive to matching personalities and providing expertise in key areas. Achitecting teams have been compared with surgical teams: a chief surgeon, an anesthesiologist, a head nurse, and one or two specialists depending upon the surgery to be performed. [BR 95 27] Effective teams can vary in size from one to perhaps a half dozen over time. Based on experience, the first member should be the chief architect, of course. But subsequent additions should be, respectively, another architect (more inventive), still another architect (different domain), and then what might be called a "field marshall" or hard-driving, mission-focused manager. [GE 92 14 and CU 92] A "large" team is considered more than four.

The composition of an effective team is remarkably close to that of the initial organization of a high-tech start-up company. The organizing principle of both is one of diversity in perspective but commonality in purpose. The greater the required diversity, the larger the source of such expertise needs to be, extending to hundreds in the aerospace business. Yet the team must remain small. The solution: the key people, the architects, are few in number and as permanent as personal careers will allow. Experts in other fields come into the team only temporarily and when the need arises. The architecting team for the Apollo lunar program consisted of about a half dozen key people for about 6 months, and perhaps four times that number for a week or so in a second echelon, with hundreds available on call for detailed analysis, reporting on the latest results of research, commenting on political factors, and so forth.

The relationship of reorganizing to product line architecting

Clearly, for a reorganization to add value to the customer and the supplier, there has to be value added to the products supplied. The subject of added value was covered in some detail in Chapter 2 and will not be repeated here. In brief, there is much more to added value than reduced cost. Cost may be only a secondary consideration for many, if not most, customers. And when all is said and done, value is in the eyes of the beholder. Nothing matters if the customer won't buy the product.

It is probably clear, too, that reorganizing and product-line architecting must be a matched pair. They certainly can't be incompatible. In fact, most companies might do well by having a common architecting team for both. There are many commonalities. Product-line lifetimes may last decades. Reorganizing and its accompanying trauma should be infrequent. A product line implies consistency of purpose, service, support, and customer base. So, too, with organizational structures.

However, in periods of rapid technological change, such as a sharp increase in the software content of a hardware product line, the whole line and its architecture may have to be abandoned — a terrifying prospect for those who have devoted a career to it. This kind of a change virtually mandates a change in organization. Ideally, the product line and organizational changes should be thought through together. Either can preclude the other from being successful. An ill-considered reorganization can mess up a fine product line. An improperly introduced new product line can mess up a smoothly running organization. Changing together they can produce a needed, less traumatic, successful revolution in both.

Ideally, the two changes should occur before, and not much later than, those of the competition. If the competition is more agile, if its products come to market sooner, if it rethinks its product–process architecture from scratch — as did the winners in the six eagles tale — then trying to play catchup becomes the ultimate management nightmare, a long-playing but frantic act of desperation. American automobile manufacturers and others have been in this position for far too long for their own good. Their culture and architectures, essentially the same for the last 70 years, are drastically different from those of the software field that is taking control of more and more of the hardware elements that the former has produced for so long. [CO 98 36-8]

This hardware/software cultural clash is not unique. It also can be seen in telecommunications, telemarketing, and inventory control. It may soon be threatening the aircraft industry. Similar clashes are well underway within the computer industry between central and distributed systems, general purpose and domain-specific designs, and between all-purpose software languages to different ones for different purposes.

Summary

This chapter by example shows the essence of organizational architecting — the process of asking the client questions based on insights gained from experience in various domains. Through those questions the client comes to the value judgments essential to creating a desirable and feasible architecture for the problem at hand.

Architecting a reorganization and architecting a new product line are shown to be closely related. Both have the same purpose, the creation of added value to the customer and the enterprise. Both are long term. Both require creating and working with an architecting team, a difficult and sensitive task with the exercise of responsibilities for both client and team being essential for success.

Notes

1. By Bob Spinrad, one of the executives involved as it was happening, as stated to the USC systems architecting class.
2. Stated in BR 75 in Chapter 11, but arguably refuted in BR 95 pp. 264-6. This author has changed Brook's original "plan to" to "be prepared to," which remains a worthwhile precaution against over eagerness to bet all chips on the first throw of the dice.

Part V

Stay the course or change it?

Setting the stage

Introduction

From its first pages this book has been about excellent organizations facing the challenge of global change. The book's response to the challenge is based on four premises:

1. Organizations are complex systems, people-based, but nonetheless *systems*.
2. Every system and organization has an architecture, or "structure" broadly defined, which largely determines what the system can and *cannot* do.
3. Systems architecting can be as applicable to the structural problems of organizations as it is to the problems of the hardware and software products the organizations create and support.
4. Systems architectural insights and techniques, heuristics and metaphors in particular, can be effectively used to sustain the excellence of organizations, their people, and their product lines — especially during times of global competition and unavoidable change.

The first eight chapters describe the rationale behind these premises: accommodating rapid changes in complexity, generating emergent values and the interactions that create them, working with multiple stakeholders and their diverse interests, and applying qualitative tools to nonmeasurable and irreplicable organizational problems.

Reorganization is then used in Chapter 9 as an example of architecting a stressful organizational process with few if any precedents and lacking information critical for a mathematically provable "best" solution. Using a small number of insights relevant to the process at hand, a systems architecting process was used to help the reader–client build an architecting team and to come to a satisfactory, self-generated conclusion.

Strategic planning for organizational change

Part V brings the concepts of the first four parts to the most difficult and critical of all organizational problems and the central one of this book — deciding whether or not to change an organization's present course to one that is markedly different in product lines and organizational structure. Because changing course for competitive reasons, be they civilian or military, is arguably the most risk-laden action an organization is likely to take, it occupies all of Part V and concludes the text.

Reasons for change

There are many reasons for an organization to change course. In decreasing order of disruption they include:

- A change in product line, or in client need, in traditional markets, emergent markets, corporate objectives, technological options, smarter machines, or in new materials — separately or together. (Example: a change to "smart" products and processes.)
- A change in the organizational structure of a client or competitor. (Examples: the change of mom and pop stores to superstores, a reorganization of governmental agencies.)
- A change in the availability of professional personnel. (Examples: system architects, software-literate designers, displaced immigrant scientists, women as combat personnel.)
- A change in regulations, taxes, financial rules, and so on. (Examples: tort liability, affirmative action, federal accounting standards, environmental requirements, and value added taxes.)

Of these, the latter three are generally adaptive, adjusting to a continually changing external environment. The first provides the greatest opportunity for being proactive, of being "ahead of the problem" instead of behind it, and of taking the initiative and the risks associated with it. As the saying goes:

The pioneers get the arrows. (G. Fox)
Yet, when all is said and done,
There's nothing like being the first success.
And surely,
Success is meaningless if failure is impossible.
(President Andrew Jackson 1812)

Prerequisites for considering change

Regardless of the particular reason for considering change, the first task usually is to assess core strengths and the capability of each for change; that is, what can each do well, what can each *not* do well, and why. Present core strengths certainly include facilities and financing, but more importantly they include people, teams, policies, procedures, suppliers, product lines, databases, business and technical contacts, and so on. These assets are difficult and time-consuming to create and can be even harder to re-create if prematurely reduced. Many are hard to measure or quantify.

The second task is to review the purposes and objectives of the organization to determine their present validity and utility. The most effective objectives are action-oriented; that is, they should lead to clearly understood and specific actions. For example, "make the best possible product at the least possible cost for a reasonable profit" does not help the organization's staff know what to do next. None of the terms are defined nor quantified. A much more action-oriented objective might be, "First, make a profit of X% above inflation and before taxes, without which none of the rest can be accomplished. Second, build a product valued as above average by the market and with warranty returns of less than Y% without which we lose our reputation as an excellent organization," and so on.

The third task is to re-think how product line, core culture, beliefs, and organizational architectures interrelate. These relationships strongly determine how *well* the organization *as a whole* can function. That is, not only whether it can make a profit, but also what the organization can or cannot accomplish efficiently and profitably. For example, a high-tech, creative, innovative organization may neither have, nor want, the discipline needed to produce a commodity for a profit.

And the fourth task is to evaluate the organization's capacity for internal change and for accepting external change by acquisition or merger. Later this evaluation will be essential in determining if and how specific changes can be made. For example, a tightly-knit, strongly directed organization could experience severe trauma and polarization if, in a merger, it were suddenly subordinated to other, culturally different, management. Such was definitely a factor in the General Motors purchase of Hughes Aircraft, could have been in the SAI, Inc. offer to take over The Aerospace Corporation, and was judiciously avoided in the Rockwell–Collins merger.[1]

A question of timing

It would be irresponsible to claim that every external change calls for an instant internal one. The landscape is littered with grand-sounding changes that either failed outright or took too long to realize their potential. Premature commitment to new technology is certainly one of the reasons for failures not only in major programs but in management reorganizations.

> *It can be just as fatal to change too soon as to change too late.*

At the same time, all managers and executives know that change — some time — is inevitable. Indeed, this kind of change is a natural part of growth and survival, a part of a sequence of need, satisfaction, obsolescence, and new needs. The first questions need to be not "what and how" to change, but "why and when," which is the focus of Chapter 10 and illustrated in the context of evolutionary change.

Chapter 11 addresses the subsequent questions of what and how in the context of unpredicted, immediate changes. Sharp breaks from the past have long been the ultimate challenge to excellent organizations and their survival. This chapter addresses the challenges and traumas of downsizing, phasing out of obsolescent product lines, startup quandries, mergers, acquisitions, and divestitures. Meeting them brings Part V to a close.

Notes

1. As recounted by those intimately involved: Wheelon, Axelband, Weiss, Boardman, Cattoi, and Rechtin.

chapter ten

The why and when
of architectural change

The only constant is change. Anon.

Introduction

As noted in the closing paragraphs of the introduction to Part V, the question of change is not one of whether or not a change should be made, but why, when, what, and how, in roughly that order. "Why" is important because, unless good reasons are evident to all affected stakeholders, some will invent their own, inadvertantly making changes which are antithetical to the interests of the organization as a whole. "When" is important because it affects hiring, education, facilities construction, and research and development. For all of these, because they take so much time to get ready for it, the future is much closer than those concerned with present production might think.

Again, the context is one of excellent organizations successful in doing what they do best. The organization is pursuing a steady course of meeting client needs by producing products based on successful architectures. It is generally understood in the company that obsolescence will call for retiring each of their well-architected product lines from operational service in due course. No executive nor manager is contemplating a significant change of course for at least several years. Isn't staying the course both necessary and sufficient for now?

The answer has to be a "yes and no." Yes, maintaining a steady evolution of successful product lines is necessary, but it is not sufficient over the long term. Other questions that need answers could be missed until too late. What is the present status and pace of the evolutionary course? How long might it last? When should architecting begin for the next product line and its supporting organization? The answers could be surprising. It could be later than one would think. This chapter is concerned with the prerequisites for being able to answer such questions responsibly and in time.

Organizational strengths and their accompanying weaknesses

The first step in being able to answer them is to assess present strengths. No organization is, or can be expected to be, uniformly strong. Every organizational architecture is the result of choices of features that determine how well, under given constraints, that architecture can perform now and in the future. Each feature, as described in Table 10.1, is strong in some ways and weak in others. Features presumably would be chosen that match the organization's perceived needs.

For example, an organization might choose features appropriate for a large, mature, profitmaker engaged in commercial and defense work. Companies like TRW, Ford Motor Company, IBM, and Shell Oil are typical examples. Another might choose features appropriate for a small, not-for-profit, agile organization engaged in "cost plus fixed fee" consulting and research colocated in owned facilities; for example, a foundation, university, or federally funded research and development center.

The choices clearly are interrelated. Some features will fit well with others, creating synergy and mutual support. But some will not, producing conflict, inefficiency, and internal competition. For example, the set of profitmaking, basic research, and bureaucracy features will not "fit" well because their individual basic objectives are in conflict. Profitmaking requires tight control, an anethema for basic research. It needs financial metrics, a near impossibility for bureaucracies providing a service.

Indeed, because in complex organizations all features are connected, a single mis-fit can produce deleterious consequences everywhere. Choices, therefore need to be made in sets, as Chapter 7 notes, and not at random. By the same token, introducing a single mischoice in an otherwise compatible set is almost guaranteed to produce chaos. Such a change was made in McDonnell-Douglas when a matrix structure was changed to a project hierarchy, with disruption throughout the company. Conversely, a correction of a single mis-fit may result in dramatic emergence of corporate capabilities.

Another lesson from the eagle metaphor

The final entries in Table 10.1 are the key architectural features of eagles and cormorants, which make both birds excellent in their own environments but worthless in the other's. The metaphor is used here to reemphasize the necessity of treating features as a set. Eagles can't swim and cormorants can't fly high by just changing one feature. Exchanging an eagle's claws for a cormorant's flippers, or vice versa, disables both birds, regardless of their other strengths. The metaphor also helps understand that some kinds of changes are impossible and should not even be contemplated. To change to one in which all strengths have to be discarded and completely replaced by a whole new set of capabilities in effect destroys the original organization.

Table 10.1 Strengths and weaknesses inherent in architectural features

Architectural Feature	Is a Strength in	Is a Weakness in
Matrix organization	Multiple projects, systems	Single product line innovation
	Rapid technology transfer	Loyalty to multiple clients
	Efficient use of experts	Meeting multiple schedules
	Open software architectures	Closed software architectures
Project organization	Single product lines	Systems from own products
	Close, loyal client relationships	Product vs. company loyalty
	Closed software architectures	Open software architectures
Bureaucracy	Service providers	Bounded tasks
	System management	Innovation
Profitmakers	Bounded tasks	System management
	High risk innovation	Service providers
Not-for-profit organizations	Financial objectivity	Working with profitmakers
Large mature companies	Mass product evolution	Synthesizing the unprecedented
Small startup companies	Creating the unprecedented	Staging to production level
Cultural organizations	Social and professional causes	Long-term system management
Basic research	Discovery of new knowledge	No net gain to sponsor
Technical training	Directly usable tasks	Extending the field
College education	Knowledge and understanding	Directly usable job skills
Cost + fixed fee contracts	High-risk development	Low-risk production
Firm fixed price contracts	Low-risk evolution	High-risk, high technology
Parametric cost estimating	Costing known products	Costing unprecedented ones
Central planning	Service networks	High-risk ventures
Single-product planning	Niche markets	System development
Owned facilities	Volatile economic situations	Technical changes
	Stable product lines	Geopolitical instability
Leased facilities	Engineering companies	Long-term production
Architecting firm	Concept development services	Personnel management
	Multidisciplinary conceptualization	Project management
Engineering firm	Large-scale development	Informal consulting
	Well-defined projects	Ill-structured tasks

Architectural Feature	Is a Strength in	Is a Weakness in
Eagle's long, strong wings and sharp claws	High-soaring surveillance Snatching weight while in flight	Can't swim at all Catching submerged prey
Cormorant's[a] weak wings and strong webbed feet	Wings unneeded underwater High-speed propulsion under water	Flying long distances Walking on land

[a] For nonbirders, a cormorant is a modest-sized, aquatic bird with dark plumage, webbed feet, a hooked bill, and a distensible pouch. When swimming underwater it looks remarkably like a small seal in shape, coloration, swimming speed, maneuverability, and, of course, the occasional need for air. An excellent example of evolution being controlled by the environment in which a fish, land animal, or bird must survive.

Why change if all is going well?

Figure 10.1 (a repeat of Chapter 7, Figure 7.7) is part of the answer. It shows how an excellent organization might view its own progress, past, present, and future. Eventually, everyone knows, things will plateau. Although organizations characteristically see their progress as still upward, there may be internal signs that it is time to think about why and when to go to the next level.

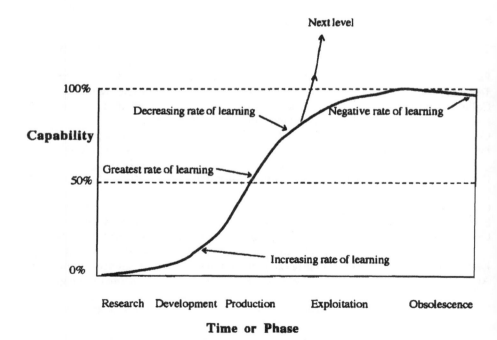

Figure 10.1 The capability "S-curve".

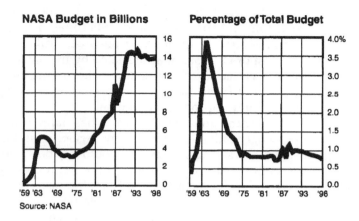

NASA Budget in Billions   Percentage of Total Budget

Source: NASA

Figure 10.2 The "public interest" perspectives.

But there is an even more foretelling *external* perspective, the one of outside observers, clients, customers, the media, and the public whose enthusiasm, interest, support, and funding, *relative to their other outside interests*, is so important. Evidence of this perspective can be seen in elections, television and radio ratings, page placement in newspapers, and especially in funding. A not atypical pair of funding curves is shown in Figure 10.2 based on NASA public documents.

The first curve (NASA budget in billions) gives the impression that NASA is doing immensely well. Funding seems to be exploding. But the second (percentage of total budget), corrected for inflation and for greatly changed funding for other needs, including a major decrease in national defense spending, tells a different story. It is one of considerable concern to NASA in the late 1990s. This phenomenon has occurred often in the past — in radars, aeronautics, nuclear energy, automobiles, astronautics, and soon enough in personal computers and software. Interest in a particular product line first increases, then peaks, and then drops to a lower relative level where it may stabilize or slowly fade away. A drop to 25% of peak is not unusual. Resurgence is possible, but rare. If it does occur, it is usually driven by different applications. Why then, might NASA change its paradigm from large, multiple instrument spacecraft to "smaller, faster, and cheaper" ones, even though it was not apparent that they were "better" in a technical sense. It was, at its core, to retain a hard-earned reputation for adventure and success and the funding priority associated with them.

A perception of continuing progress, like a perception of excellence or success, is one of the principal assets of any excellent organization. Loss of it has serious implications — fewer of the best and most innovative professionals can be recruited; new justifications and sources of support, from government funding to foundational support of basic research, no longer appear; the best and, consequently, most mobile executives begin to look elsewhere for challenge; the organization gets the reputation of being mature if not staid; and the existing and successful product becomes a cash cow, an ongoing but uninspiring source of results and revenue. Truly innovative ideas and products get lost or at least not appreciated by the outside world, in what can become a less-than-excellent setting. The originally exciting excellent organization slowly fades away. The future becomes, well, *boring* to all but a dedicated few and unproductive to them.

This decline is by no means unique to NASA. It has happened often to the military forces, passenger railways, government service of all kinds, TV and radio programs, and marketing services. It would not be surprising if it occurred sometime in the near future to e-mail, Internet browsing, sport utility vehicles, and other developments presently on the rise.

Whatever the cause, public perceptions of the relative worth of excitement and excellence somehow get lost. It's more than enough to cause concern. The reaction, usually unsuccessful, of some organizations is to change their nature and purpose, to see the grass as greener on the other side of the fence, to go and compete on somebody else's turf. Go for discount sales. Catch terrorists. Make better chocolate chip cookies. Whatever. Such enthusiasm for change ignores the fact that the present owners of the turf are formidable competitors, highly skilled and highly incentivized to do what it takes, including changing themselves, to keep their turf — especially if their markets are already over-supplied.

Although it originated for a different purpose, the insight holds once again:

> *Never leave until you know where you are going —*
> and know it well!

In other words, if you haven't at least a corner of the turf, don't try to invade someone else's. A better strategy might be to restate the question of changing-to-survive not in terms of a solution (e.g., go commercial, etc.) but of the problem.

> *Don't assume that the original statement of the problem — much*
> *less the solution — is necessarily the best, or even the right one.*

The subtitle of this book, *Why eagles can't swim*, could imply that the eagles' problem is to learn to swim. It isn't, of course. It is to catch fish which, in turn, is an important part of the even more basic problem of survival. The "swimming eagle" metaphor was deliberately misleading. It was chosen to illustrate the very common error of jumping to a conclusion of the "best" solution before the real problem is defined.

Asking the right question

A famous example of effective technical use of the *"don't assume"* insight has been provided by Harry Hillaker, the architect of the F-16 air-to-air fighter.

The original U.S. Air Force requirement was for a supersonic fighter to provide a swift exit from a combat zone once its mission had been completed. But, with modern air-to-air missile technology being what it is, a well-launched enemy missile can easily catch a single departing aircraft. Speed does not equate to safety. Maneuverability does. Paired aircraft flight does. When the problem was restated as evasion during escape, the answer became an F-16 so maneuverable that it could evade even a pursuing missile attack. It had been *assumed* that the solution was supersonic flight.

A commercial example was the profit strategy of David Packard at Hewlett-Packard. Most people assume that profit is the difference between selling price and cost; hence, the effort to increase the price and decrease the cost. In the extreme, this assumption leads to monopoly (or market share) production of poor products. Packard made a different assumption, that profit was the "tip" provided by the customer for extra service (worth) to that customer; hence, better products that increase that worth. In a way, the customer was sharing the additional value that Hewlett-Packard had enabled that customer to attain by the customer's use. As addressed in Chapter 3, sharing added value with one's supplier (i.e., Hewlett-Packard) is sound business for both.

A second assumption and for years a valid one was made by the people of the Bell Laboratories; namely, that they were indefinitely chartered to develop the world's best telephone communication system for all individuals willing to pay the price per call. That assumption was demolished by court order to be replaced with the assumption that their research was intended to benefit a competitive, profitable, company that would charge primarily for access to its facilities independent of actual use. Another, almost fatal, assumption by the Bell System was that the architecture of the system itself would remain much the same — ground microwave and cable linkages through a network of central stations. As such, billing could be based on the distance between the users. When satellites were shown to be possible in the late 1950s, J.R. Pierce of the Bell Laboratories proclaimed:

> *"We must not be the American Association of Railroads*
> *in the era of airplanes and trucks!*
> *We must be in the communication satellite business if we are to be*
> *communicators and not ground facility owners!"*

Little understood at the time were some of the many consequences of that warning. Technically, for example, satellites would provide a completely different architectural configuration, one for which distance on the surface of the Earth was immaterial. Any two points in view of the same satellite could communicate directly, which for synchronous communications satellites effectively meant any two points on the same side of the Earth for the same cost, and without the use of any Bell System long line links at all!

Forty years later the telephone system architecture is changing once more. No longer will telephone numbers be associated with specific locations but with specific users, regardless of their location on land, sea, or air, moving or not. Only slightly less convenient and somewhat more expensive than before, the required "smart" architecture upsets all prior assumptions of what personal communications means in billing, connecting, supplying, and operating the telephone system.

Hence, given a presumed statement of the problem, the first step is not to press forward by looking for feasible solutions, but challenge it! It is to go back in time — in the "purpose" direction — to determine the so-called problem's underlying purpose; or, as Gerald Nadler [NA 81] puts it, "The purpose behind the purpose behind the purpose..."

> *No organization can live without a viable purpose —*
> so find it and only *then* design accordingly.

Only then can architecting parameters begin to be prioritized that have imperatives that must be met regardless of the final narrower application. It is at this level that architecting is most valuable.

To be useful as a practical guideline for action, this insight demands more than a rote answer. It is not enough for architecting purposes just to respond, "to survive" because that leaves unanswered the more difficult questions of who or what is supposed to survive. The present employees and managers? The product line ? The stock holders? The facilities?

It is not enough to answer by saying "times are a-changin' and so must we," or that "the competitor is doing so and we must follow." The first statement leads to random changes for the sake of change; the second to mindlessly following a competitor down what later might become a mistaken course. Remember the American chase after a Japanese-like culture to "correct" American work habits, a perceived need that was proved baseless by green field transplants?[1]

In the NASA example, the decision was, indeed, to stay the course in its most important objective, its originating purpose — to explore space for the good of mankind. The mechanism by which this would be accomplished, manned and/or robotic, was left open.

A rationale for staying or changing the course

Architecturally speaking, the basic rationale behind changing or staying the course is to maintain or increase the organization's value as perceived by the clients, other stakeholders, and, yes, itself.

If so, then the original question of "why change" might be answered not by "to survive" but by finding a marketable purpose and a feasible way of adding future value to it. "Survive" will then be a consequence and not a reason for adding marketable value.

Note that "adding *marketable* future value" is not the same as adding the most revenue, profit, or other financial metric to the company's books. The former is about client interests, the latter largely stockholder interests.

When to change

There is a famous insight which gives the three most important factors in management change as

Timing. Timing. Timing.

Change too soon and the necesssary prerequisites are rushed or absent. Success, if any, is postponed. Stakeholder support needs continual reinforcement, not premature promises. Employee morale and support slackens under "tomorrow, surely" strategies. Management understanding of possible consequences is less solid. Facility availability may not exist. Customer reaction is likely to be ambiguous. Even worse, a series of failures to deliver as promised can ruin a company's reputation.

Too late and the competition may have preempted the field and skimmed the most profitable sectors of it. After all, coming in second teaches that:

There's nothing like the first success!

and probably nothing worse than having the mountain produce a mouse, of promising the moon and delivering a pizza, of hyping what turns out to be a trivial modification of an existing product.

Beware of evolutions disguised as revolutions.
More often than not, a new concept is another component in an
old box, not another tool box. (Gold, Jeff 91)

A rationale for deciding on timing

One of the best-known sequences for initiating action is detection, identification, decision, action, and assessment — repeated over and over. First, detect that some kind of problem, internal or external, exists. Second, determinine its cause or causes as accurately as possible Third, decide to act but only at an appropriate time.[2] Fourth, act precisely and with will and determination. Fifth, assess the results for effectiveness and then on decide what to do next. In organizational terms this means:

- Detect that something is amiss by observing that factors that have been doing so well seem to be changing, even if the causes are unknown.
- Determinine or diagnose the problem as soon as possible, especially if two or more factors seem to be behaving strangely. Single factor anomalies can have many causes. Two or more concurrent anomalies sharply limit the number of possible causes and help to determine the urgency of corrrecting them. For example, decreasing profits can be from at least two sources, decreasing revenues or increasing costs. Which one is the culprit? As insights:

> *Knowing that an anomaly exists can be more important*
> *than the actual anomaly.* [Kjos, K. 88]
>   and
> *Correction of an anomaly is not complete until a specific*
> *mechanism* and no other *has been shown to be the cause.*

- Take what appears to be the best action.
- Assess the results to see if the problem is fixed.
- Decide, based on the assessment, what to do next.

The most critical of the five steps is probably the second one. If the diagnosis is incorrect, the subsequent steps can make the situation worse. Look for and detect changes in early warning indicators, requiring a close connection with the prior step if both are to be effective. The most serious threats and the strongest incentives to change undoubtedy are those affecting the main product line(s) such as any decrease, absolute or relative, in any performance factor. In the NASA example, the first time would have been in 1965[3] if not earlier. Other indicators in other contexts include quality, cost effectiveness, delivery schedules, warranty costs, consumer dissatisfaction, software technical support, subcontractor responsiveness, marketplace restructuring, discount pricing, and so on.

Whether the diagnosis predicts product-line death, abandonment, evolution, or major re-structuring, re-architecting, or re-inventing, depends strongly on three factors:

1. Is the product line still economically viable; that is, are there consistent answers to the questions: *Who benefits? Who pays? Who supplies? Who loses?*
2. Is it still adding value commensurate with cost and is this likely to continue for some years?
3. Can the present architecure continue to add still more value?

If any answer to any of the above questions is "no," the organization, no matter how well it seems to be doing at the present, can be in serious architectural trouble. If *all* are "no," the situation is approaching disaster for the organization. It may have run out of time.

If all the answers are "yes," the followup question might well be, "How soon do you expect any of these answers to be 'no?'" If the answer is within 5 to 10 years, it is time to start thinking hard about major change. NASA's "smaller, faster, cheaper" strategy started a few years ago is clearly an effort in that direction. Judging from the answers the diverse NASA stakeholders might give to the questions addressed earlier, it was none too soon. Stay tuned.

> *Generally speaking, achieving major change takes about 10 years*
> assuming *few intractable internal or external barriers.*
> *Otherwise, to the great disappointment of its advocates,*
> *it may take at least a human generation.*

The surprise in this insight is not that change takes so long but that it can be so short! The sources of this heuristic are remarkably broad. The 10-year period has been observed in such dissimilar situations as cultural change in nonprofit organizations, technological base changes in defense system suppliers, affirmative action in private firms, the strategic thinking of executives in major corporations, discipline changes in universities and colleges, regulatory changes in public utilities, and deregulation throughout government and industry. [REN 95 8] [DA 94 251] The 10-year period, even so, applies almost solely to relatively receptive, strongly led, seriously threatened organizations. Those in the information systems sector are prime examples. Rapid changes in technology understandably are very strong factors in that sector.

A different, 20 to 40 year, generation-long period applies in other situations, such as "glass ceilings" and "minority parity" in which skilled and experienced white males now in place have to leave, retire, or die before major change becomes practical. And while the change process is underway, the organization must continue to be excellent. There is no "time out for change." As Tushman, et al. expressed it in 1988: [TU 88 102-130]

> *"The most successful firms both manage incremental change*
> *and are able to implement strategic change*
> prior *to experiencing (in anticipation of) performance declines.*

Assessing organizational architectures

Whether incremental change can accomplish its purposes or not depends on the long-term robustness and reliabilty (absence of defects) of the structures of the product line and its supporting organization. A sampling of the better insights for assessing these qualities is: [RE 97 241-2, restated unless otherwise indicated]

- *The test of a good architecture is that it will last. The sound architecture is an enduring pattern.* (Spinrad, Robert, 1993)
- *A good organizational structure has benefits in more than one area. As civil architects would say, the structure has" good bones."*
- *The bitterness of poor performance remains long after the sweetness of low cost and prompt delivery are forgotten.* (Lim, Jerry 94)
- *Before-the-fact architecting is much to be preferred over after-the-fact diagnosis.*
- *The first quick analyses are often wrong.*
- *Unless there is an effective internal communication system so that everyone who needs to know does know, somebody, somewhere, will foul up.*
- *In correcting organizational problems it is important that all the participants know not only what happened and how it happened, but why, as well.*
- *Chances for recovery from a single management failure or flaw, even with complex consequences, are fairly good. Recovery from two or more* independent *but coincident failures is unlikely in real time and uncertain in any case.*

Organizational modifications based on architectural assessments

In point of fact, history shows that very few organizations undergo more than one or two wrenching changes in direction and still survive as the same organization with the same purposes. Most, however, can and do manage incremental changes over a period of decades, given the opportunity and context to do so. Partly, of course, the reason is that changing course both radically and successfully is so difficult.

It is also true that truly successful, long-term, product-line architectures last for decades and, hence, opportunities for radical changes are infrequent. Another reason is that organizations merge with others, divest key elements, or otherwise lose their unique character and culture. Little may remain but a then-meaningless company name or a set of uninformative letters. For purposes here, the organization as defined by its purpose and its added value would no longer exist. No wonder there are few successful radical changes. One of the hopes of this book is to increase their number by suggesting appropriate insights and prerequisites to those that can use them.

Of those that do stay the course in function and purpose, most began with fine long-term product lines or services, continuing for many decades to evolve their products as evolutionary systems.[4] Characteristically, these have been organizations with large capital investments amortized and improved over many years. Public utilities (power, telephone, highways, and airways), defense installations, and public universities[5] are examples. Their original purposes remain much the same, their architectures, usually stable, seldom change radically (but they *do* change), but for almost all their lifetimes their changes are incremental and partial. It is just too *expensive* to change everything in any short time. Even major wars and technological advances can do little to them but start trends in new directions. As to be expected, their structure, policies, procedures, and governance have developed in ways that reinforce this stability. If and when they change, it is likely to be due to very strong external pressures — antitrust suits, deregulation, changed teaching standards, ethics concerns, political and ethnic demands, and so on.

Annual state of the organization reports and 10-year plans

A familiar challenge posed by many managers is, "Why plan? We don't even know what tomorrow will bring. Just be flexible! Stay loose! Why plan? " A straightforward answer is, "So that present decisions will continue to take the future into account." Unless that happens, options can be missed, preferred directions can be precluded, and present weaknesses may not be corrected in time for future change.

A pair of insights can help minimize such losses.
The first recognizes that:
*Unless you know where you are, you can't know
where you are going.*

This insight is the reason for annual status reports, enshrined in the U.S. Constitution of the by the requirement that the President present, annually, a State of the Union report to the Congress.

The second insight is a companion to the first:

If you don't know where you are going, you won't get there.

The manager's query is certainly valid. The future *is* unpredictable in any detail. But that does not mean that the future can't be affected to advantage. And there are many reasons to work toward a *preferred and feasible* future. Annual 10-year plans are provisional visions of how to get there. Their scope, preferably, is broad and includes both product-line and organizational architectures. The time frame is far enough in the future to avoid the volatility of day-to-day events, but not so far as to lack credibility.

Long-range planning: the NASA–JPL Deep Space Network

As an example of long-range planning, 10 years (1957 to 1967) of annual 10-year plans for the NASA/JPL Deep Space Network (DSN) set the direction not only for its first 10 years but for the next 10 as well. They were remarkably prescient technically, useful organizationally, but only modestly predictive managerially. Among the missed managerial predictions, the director of the DSN (and author of the first plans) was recruited by the Department of Defense just as the first 10 DSN years were ending. As it happened, the second 10 years in NASA, as implied in Figure 10.2, were quite different from the first 10 and the second director was indeed a better match for them. The DSN professional staff is justifiably proud that the DSN, over a 40-year time span, was improved by a factor of about a million, went through at least four block changes, served a half dozen countries and dozens of flights, and communicated with and positioned spacecraft to the edge of the solar system 10 billion miles away, all without jeopardizing flight mission success, all on demand at any time 24 hours a day. Even more astonishing, as an organization, the DSN was able to commit such steadily advancing performance to the nearest 10% continually more than a dozen years in the future. It knew where it was and where it was going.

Both status reports and future projections can be made still more helpful by comparing each year's performance with others. Trends and correlations become more apparent. Flaws in assumptions are more easily identified. The resistance to, or successes of, organizational changes can be better appreciated. They also may help estimate the evolution and maturity of the organization's products, alerting it to potential abandonment at some point in the future. With careful review they may even keep reminding their eagles that future change is not about learning how to swim but how better to catch fish when conditions and needs change.

Summary

To change or not to change, is that really the question? Probably not. Change, sooner or later, is inevitable. Instead, the need is to find a course change which will add future value as perceived by the client in particular and stakeholders in general. To do so takes time, typically 5 to 10 years for major organizations still in excellent shape, which means early detection, prompt and accurate diagnosis, decisive action, careful assessment of the results, and change as needed all *prior* to performance decline. Any change, evolutionary or radical, takes long preparation. In any case, excellent organizations, in particular, should probably avoid changing to any course in which its presently essential strengths would become unavoidable future weaknesses.

Two tools, annual state of the organization reports and 10-year plans, help assess the status and future of organizations undergoing evolutionary change. A succession of such annual documents can help them understand their need and timing of product line retirement and phase out.

Notes

1. See Chapter 2, pages 2-10.
2. One of the more recent technological additions to the defenses against a siege has been the use of sensors distributed liberally around the besieged area to detect the approach of besiegers. The temptation was to shoot at anything the sensors detected. As military intelligence experts knew, that was precisely wrong. *Don't* respond! Not yet! First, to do so gives away the location of the sensors. Second, it is much more effective to use the sensors to determine enemy intentions to gather for attack. Then, just when the enemy forces are concentrated to do so, launch a withering artillery fire at the concentrated forces — *not* at the sensors!
3. In 1960, the Congress committed itself to fund President Kennedy's challenge to go to the moon. It also, in its accompanying reports, explictly did not commit to an ongoing manned flight program — and *still* has not. NASA administrators of the time knew this, but were either unable or uninterested in forcing the issue, one which returned to plague them in 1970 when they fought to build the Shuttle and a decade later to build the space station. The stuttering and stumbling for the next decade is graphically shown in Figure 10.2, with the present trend again downward.
4. See BE 89, Beam, Walter H. "Evolutionary Systems: Arguments for Altered Processes and Practices," The Beam Group, Alexandria, VA, ca 1990, and *Command, Control, and Communications*, 1989, same author. Beam, in effect,

assumes the existence of an organizational and product architecture; that is, that major architecting has already taken place and the remaining problems are those of more effective evolution.

5. Professor Steve Sample, President of the University of Southern California, has pointed out that the time constant ($1/e$) of a major university is about 100 years. Many of them, all over the world, have long since celebrated their centennial, even bicentennial anniversaries. Their objectives have remained much the same but the nature and relative importance of departments and disciplines have varied wildly — or at least as viewed with a 100-year perspective.

chapter eleven

The what and how of radical change

Expect the unexpected. (Butterworth, William 1991)
But, isn't that a contradiction in terms?
Not if the only constant is change.

Introduction

The previous chapter concerned the inevitability of change and the ways of maintaining the change process as a stable evolution. The strategies for doing so have been developed after generations of trial and error. By this time they are established and successful in both the public and private sectors. Namely, reinforce the strengths and minimize the weaknesses. Steadily improve the organizational support of the most successful product lines. Detect if and when something is amiss in one or more of the key financial indicators. Determine the likely causes of the anomaly, act in a timely manner to eliminate them, and assess the results. Repeat until the evolutionary process is again stable. Prepare, at the end of every fiscal year, status reports of the present condition and 10-year plans for the future. These are the future-oriented strategies for constant change, essential for evolutionary product lines and multifaceted companies.

Two examples of private, multibillion dollar organizations that long have been based on such strategies are the Hewlett-Packard Company, an instrument and computer manufacturer, and Lowe Enterprises, a real estate development firm based in Los Angeles. They share no less than nine defining principles: (1) a financial structure that allows them to survive the unexpected, (2) a diversified and multiasset base, (3) few or no stockholders other than members of the owning family, (4) decentralized management of major sectors, (5) continuity of their professional staffs, and (6) a skill at timing the startup and phase out of their products or their real estate holdings, respectively. To these six should be added three more: (7) both consciously and successfully made themselves virtually independent of wide swings in the national and global economies and their stock markets by using these principles as a closely coupled *set*, or system; (8) each has occasionally strayed from the set, regretted it, returned to it, and prospered; and (9) Both were founded by strong leaders, David Packard and William Hewlett of H-P, and Robert J. Lowe of Lowe Enterprises, who led their organizations and rigorously enforced these principles for many decades. [FU 99]

Future-oriented strategies clearly can work well for an organization with a well-architected, evolving product line or asset base and a corporate structure created to support it. If there is any serious risk to such strategies, it is one of complacency and self-absorption — of taking one's professional expertise for granted, of increasingly looking inward for efficiency to the exclusion of looking outward to effectiveness, and of assuming continued external acceptance and support. A lesser risk, certainly for the two companies above, is an effort to be excellent and successful in other than their own fields.

This chapter recognizes that these stable evolutionary strategies can be abruptly upset by external events, ones that mandate completely new products and, hence, new ways of managing their development and sale. These external events, while there might be precursors, seem to come upon the scene suddenly, as if triggered: wars, new environmental regulations, global financial crises or bonanzas, stock market shifts or arrivals on the market of a competitor's products that "overnight" make the present product line obsolete. All can tear an organization apart if the company is caught unaware by being bought, sold, redirected, merged, dismantled, or moved. On the other hand, the company also could be suddenly infused with massive capital, rejuvenated in another guise, teamed with an ideal partner, or blessed with a huge and unexpected latent market.

Strategically upsetting events, in any direction, invariably force "bet your company" decisions, decisions which test management's resolve never to compromise the company's primary goals. (Ben Bauermeister 1998) Can it, or should it, continue to maintain its openness (or tight security), trust (or isolation), and freedom of expression (or discipline), or to serve national security (and/or) protect the environment? The decision to change may involve divesting of a major strength such as GE appliances, IBM software, General Dynamics defense systems, or H-P systems contracts, in order to use its resources for the course change.

Events like these can mandate major, internal, architectural restructuring — going from hardware-intensive to software-intensive product lines, from a hand-crafted product to an automated production line, from matrix to project management, from a not-for-profit to a for-profit status, from being an entrepreneurial startup to becoming a major supplier, from being a defense contractor to an international conglomerate, or from being a regional power to a global one. Every relationship between both internal and external units can be affected. The relative importance of core strengths will have to be judged anew. Individuals will have to rethink their hopes and fears. Previous plans based on orderly change will be jolted. Excellence will have to be re-established and respect re-earned. Success can no longer be taken for granted.

For the sake of our thesis, consider these changes to be irreversible, urgent, and for reasons never even imagined in previous 10-year plans. These are *radical* changes and, in today's global world, they occur surprisingly often. The list of widely respected excellent organizations that have simply "disappeared" within a very few years after the initial shock keeps growing. But so does the list grow of new companies in product lines that, unheard of a few years ago, seem to have become absolute necessities, only to fade in as short a time. All excellent, some succeed and others fail. Trying to understand why and what else might have been done architecturally to modify the results has brought this book into being. This final chapter is intended to help in responding to the shocks of downsizing, phasing out of obsolescent product lines, startup quandries, mergers, acquisitions, and divestitures.

The single most important response to all of these shocks is to be prepared for them. That seeming contradiction begins the chapter. To some extent the most effective preparation depends upon the nature of the shock, of course, and by definition that nature is unknown ahead of time. As it turns out, however, the same preparation is effective for many of the shocks so, in the interest of brevity of text, the most common ones are introduced first and the more difficult ones later on. This order happens to match the increasingly greater trauma of the examples, from downsizing (usually anticipated) to divestiture (as traumatic as a divorce).

Preparation well in advance

The first and arguably most important strategy for handling unexpected radical change is to be prepared for it. On the face of it, that statement may sound nonsensical. It is, of course, more difficult to prepare for a divestiture than for a 10% downsizing, but it is not impossible. Arguably, the further in advance, the more effective the preparation.

Certainly the initial shock can be lessened. And almost any preparation is better than none because visible preparation is an acknowledgment that change is not only possible, it is probable. Visible preparation is of enormous psychological value when the unexpected event finally does occur. It will then be a "so, what else is new?" instead of being speechless, helpless, and dismayed.

Responding to event shock

> Sometimes how something is done is more important
> *than* what *is done.*
> *The certainty of misery is better than*
> *the misery of uncertainty.* (Pogo)

Managers and other professionals who have gone through radical changes are the first to urge immediate action by the leadership to acknowledge the

fact of the imminent change and to assure all stakeholders, from past sup-
porters to the newest recruits, that they will be kept informed as further
events transpire. If appropriate and accurate, leadership should describe the
situation as one of unanticipated challenges and the opportunities. If that is
not and cannot be true, then at least be honest!

To do otherwise at such a time, to remain reclusive and silent, can
paralyze an organization when it most needs to be vigorous and responsive.
If left ignored, the best and consequently most mobile of the professional
staff will leave, diminishing the value of the organization as a whole. In
short, the first moves must be to buy time and to retain the trust and
confidence of the staff until much more is known and negotiated.

Downsizing

In the mid-1980s as the Soviet Union was being torn down, it was evident
that the U.S. defense budget would have to decrease and, after 50 years of
Cold War, the prospect was daunting. As with other defense-oriented companies, The Aerospace Corporation knew it would not be immune from
funding reduction, though when and by how much was anyone's guess.
Nonetheless, preparations were made for the eventuality, beginning with a
study that showed that the organization could at any time, without its
mission being compromised, take a 10% per year reduction for at least
3 years, given only an 18 month notice, *providing that* long-term plans for
facilities, hiring, retirement, debt structuring, and the use of fee were modified well in advance, which of course they were. More by coincidence than
planning, articles were written by its CEO for its company paper, *Orbiter*,
reinforcing the culture and beliefs of the company — fundamentals that the
company could not lose and survive.[1]

The Berlin Wall came down. Funding went down year after year. The
company morale was bruised but not broken. A decade later the storm had
passed and the company was still standing, willing and able to support its
40-year mission of national security. Was all peace and quiet during the
storm? Of course not. The political process of budget reduction was accompanied by both dedication and betrayal, but the reasons for them, as Brenda
Forman's insights [FO 97] first described in 1989, were at least *understood!*

The Aerospace Corporation was not alone. The defense industry (that
much-maligned military-industrial complex), in an astonishing exhibition of
realistic moves and Defense Department encouragement, literally reduced
itself. Affected individuals and cities adjusted with far less trauma than had
the U.S. automotive industry 10 years before. Amazingly, there was little
bitterness, recrimination, or social upheaval when companies disappeared
in mergers, when the military forces were sharply reduced, and when "new
starts" were cancelled. No major political figure raised the call for jobs, jobs,
jobs. The changes were understood and acknowledged as necessary. A wise
policy was announced as the downsizing began that no decisions would be

made that could not be reversed. As reductions continued, appropriate contract modifications were made that encouraged industry to consolidate in an orderly, even profitable, manner. As an example of a well-managed radical change, that restructuring has had few peers.

Phasing out of obsolescent product lines

Not all radical changes have to be of such a gigantic scale nor involve so many people as downsizing to be very serious to those directly involved. The need for preparation well in advance can be just as important in handling the changes caused by product and process obsolescenses and eventual phase outs in Chapter 7 (Figure 7.7) and waning public enthusiasms in Chapter 10 (Figure 10.1). Obsolescence should be planned, according to this strategy, just when the slope is greatest; that is, just when the greatest progress is being made. This can be very hard to do in the enthusiasm of the moment. Nonetheless, it is certainly better to plan ahead when things are going very well, than after an unmistakable observation that they have gone stale. As was noted earlier, change takes time which can only be bought ahead of time, and never thereafter.

The story of the six eagles, in which excellent companies sequentially lost contracts, demonstrated the need to understand one's own limitations in trying to introduce new achitectures. The lean production story showed that lack of understanding of systems fundamentals led to piecemeal attempts by the American automotive industry that failed to deliver the across-the-board emergent capabilities that the Japanese had already achieved and openly shared.

The need for consensus

Of all the factors likely to cause trouble despite careful thought beforehand, the one that no doubt ranks first is a failure to prepare stakeholders for an imminent change in their company's product line. Although the responsibility for this task is hard to place, most observers would lay it at the feet of the chief executive officer. For small companies, that placement is no doubt correct, proper, and practical. For large companies, however, few CEOs have the training or time to take it on — *especially* during periods of radical change — much less of achieving consensus on why, when, what, and how.

Historically, facilitating stakehholder consensus has long been an important part of architecting. The tools, as have been discussed, are the ability to question, listen, and understand; the skill to use insights, metaphors, and models; and an extra degree of tact and patience. It helps to be charismatic and visionary, of course, but it is essential that the architect be a communicator continually maintaining the essentials of the original architecture against the hubbub of implementation problems in the background. Three systems architecting insights explictly target architects on this issue:

There is no such thing as immaculate communication!
So,
Don't ever stop talking about the system —
or a change underway. [LO 89]

Listen and remember that communication by definition
is bi-directional. (Art Collins, ca 1950, courtesy of Bob Cattoi)

Recognizing the need for achieving and maintaining consensus of stakeholders on the "vision," a number of top companies have established chief architect (or engineer) positions for structuring of product lines and, to a limited degree, of the organizations to support them. It is a difficult role. It is simply too easy for other professionals, from CEO to assemblers on the production line, to concentrate on doing their own jobs and to forget the "big picture" vision back of the product and its organization. The architect must, therefore, protect the vision by talking, talking, and talking about it, because:

Unless everyone who needs to know does know,
somebody, somewhere will (probably inadvertently) foul up.

Almost needless to say, consensus in many areas must have been attained well before a radical change is imminent. Attempting to answer such questions as "who are we" in the middle of a merger is clearly much too late!

The startup product and organization

The three phases of innovation are product orientation,
profit orientation, and process orientation.
(Bauermeister, Benjamin 1992)

There is a standing myth among enthusiastic would-be founders of companies that if the startup product is an early success then expansion, fame, and fortune is guaranteed. In fact, most of the time it is not; not because of the product, but because the transitions between phases are originally unpredictable, later unplanned, and therefore disruptive. The founders can become so focused on the innovation phase that they forget that unless the product is engineered to be producible profitably, it will die. They can be so taken with the early profits that they fail to produce enough fast enough to stave off the competition galloping along behind. Even if all these issues were thought through in the beginning, few startup entrepreneurs are ready with a follow-on product, having not invested in its development soon enough to again keep the competition at bay. More about this problem later.

As can be seen, the truly difficult times for the entrepreneurs are the transitions between the three differently oriented phases. Getting started and convincing others of the desirability and feasibility of the product requires someone inspired, talented, and creative. Disciplined profit-making demands someone process-oriented, an outlook seldom satisfying for the inspirational founder. But the most difficult transition occurs when the product goes from being an inspired, high-tech wonder to being a run-of-the-mill commodity, something others can grind out once shown the way. Unquestionably going from one phase to another is very difficult for any founder, even with the best of intent. Commodity production efficiencies may be so far removed from the original motivations of the founders that their Board of Directors may have to remove them from their own company to save it. Tragically, the ousting of the original innovators may destroy the company's identity.

To compound the trauma, the faster the external world accepts and then demands the product, the faster these management steps must be taken and the less the time available to prepare for them. Initial investors may want to cash in and move on. New professionals, the so-called "bean counters," may move in. Unexpectedly large facilities may have to be built with limited cash on hand to pay for them. Government buyers may arrive armed with stacks of specifications and regulations.

Change then can become uncontrollable if the founders or sponsors are not prepared to handle the transitions properly. In the worst case, a highly successful initial development may be dropped because the parent organization may not know how to either incorporate it, establish a wholly owned subsidiary, license it, or lock it up for future use — all strategies which have been used successfully in the past.

Xerox, before its recent transformation to The Document Company, had the unenviable reputation both of having one of the finest research institutions in the world — the Palo Alto Research Center (PARC) which produced a steady stream of remarkable product opportunities — and of being unable to exploit its results. The best known of these opportunties, one largely funded by the Defense Advanced Projects Agency (DARPA), was the architecture of the desktop computer, which was instead adopted and built by Apple Computer, Inc. as the very successful Apple and Macintosh PCs.

In all fairness, Xerox was not the only excellent organization unable to exploit its own ideas. Like many other companies, Xerox was run at the time by "the old-timers" or what at Xerox were called "the copier people;" executives who had made Xerox a household word and made their decisions based on that success. Just like the six eagles. Even as its commanding share of the market dropped alarmingly in a flood of competitive copiers, it still was difficult to convert Xerox to "The Document Company." [SP 92]

Spontaneous inspiration?

One of the misconceptions about innovation is that it is spontaneous — the *eureka!* phenomenon, as it is often called. Possible but rare. Reaching the point of inspiration usually takes months if not years of immersion in the field, much thought, sleepless nights wrestling with some subapplication of the idea, and only then an intuitive "flash" of realization of its importance. Thomas Edison expressed it as "invention is 99% perspiration and 1% inspiration." Jonathan Losk expresses it in management terms with a double-meaning insight:

> *Inspiration is a function of preparation.*
> *Preparation is a function of purpose.*
> *Thus,* prepare to be inspired. [LO 89]

The first meaning of the underlined admonition applies to the innovators, themselves. It calls for adequate time and preparation after a purpose or problem has been identified but before the inspired solution becomes "obvious." Jumping to a too-quick solution, with competition all too eager to do better, can be fatal.

The second meaning of Losk's insight applies to the management of a startup organization and its quandry of "what next" after the success of the inspired first product. The insight calls for starting the second and even third product, which are going to take time to be conceived, very soon after the first one reaches the prototype stage.

Yet even that is not enough to avoid an early demise. Innovation not only takes time, it takes practice. Practice comes from prior task assignments in a supportive environment, whether the context is a family with children or a professional in a supportive company. One such company is Hewlett-Packard in which continual innovation supported by astute financial policies has been, from the beginnings of the company, a cornerstone of its success. In contrast with "commodity" companies that invest less than 1% of their revenues in innovation, H-P by policy invests 10% and, comparably important, makes no demands that the projects justify themselves before beginning. Less than half of its off-the-shelf products are more than 3 years old; that is, more than half of its over $10 billion per year of products has been conceived and produced every 3 years. Change is built into everyday practice. In such organizations, one more "startup" or quick demise is no terror, it is an indication that all is well.

Mergers, acquisitions, and divestitures

Three of the most traumatic, event-driven, and sensitive changes that can affect an organization are a merger, an acquisition, or a divestiture. For brevity in words and because the consequences to an organization are similarly wrenching in all three, "mergers" will be used as the surrogate for all three.

There are many reasons for the rapidity and secrecy of the merger process, but the one topping the list is likely to be financial. Because of the fluctuation of a company's perceived value on the stock market, the timing of a merger announcement, even to the nearest stock market day, can be critical. A few days later or earlier and the merger prospect might never have seen the light of day. No wonder a merger is an event shock to the people most directly involved.

Preparing for possible negotiations to merge

Very few mergers occur without a fair amount of negotiation beforehand. It behooves both parties to be prepared for them. Many of the strategies already mentioned for management of evolutionary change can be of particular value in that preparation.

An up-to-date, annual status report and a series of annual 10-year plans can provide vital information on just how difficult, or easy, the change is likely to be.

Knowing one's culture, beliefs, motivations, organizational purposes and identifying the perceived strategy of present and potential competitors (above all, knowing what both prior and future clients value most) can make subsequent architecting far easier and more efficient.

It helps when the negotiations become difficult if "nonnegotiable" imperatives can be established well before any event shock, especially if some imperatives have to be traded away later. It is considerably more convincing to be able to say that these imperatives are "long-standing" and not just invented for the occasion.

What should not be changed

Ideally, everything possible should be on the table for negotiation in the interest of establishing fit, balance, and compromise on key capabilities of the final merged organization. It is generally agreed, however, that maintaining at least some aspect of the original corporate identity helps the merger process come to closure more quickly. Organization architects can assist, almost uniquely, in achieving solutions appropriate for enterprise success and stakeholder interests.

For example, continued use of company names, trademarks and family names, like Collins, Lockheed, Allen Bradley, Daimler-Benz, and Nissan, and company logos like the Aerospace "A" inside a circle, the italicized *TRW*, and the "Big Blue" of IBM, can provide rallying flags useful in maintaining company identity, loyalty, and dedication during and after their merger.

In general,

> *When implementing a change, keep some elements constant*
> *to provide an anchor point for people to cling to in time of stress.*
> (Jeffrey H. Schmidt, 1993)

It was just such a realization that led to the articles on belief and culture in the Aerospace *Orbiter*. Without going into the specifics of that case, it was clear that the people of the company did not want to abandon their mission of national security, their hard-earned credibility and reputation for objectivity, nor their opportunity to take on challenging system-level work — even if that meant that the organization had to become significantly smaller. These were nonnegotiable anchors, firm choices no matter what the nature of the radical change.

High-level interfacing

From a systems point of view, the most important issues to be negotiated in a merger are those involving the interfaces between the parties because the interfaces are the source of the added value of the merger as a whole. However, as Gacek's research implies, if the merging organizations are structurally different, they are likely to be incompatible,[2] and the merger will fail to deliver that value. One possible solution comes from the use of an insight:

Compatibility is easier to achieve in the abstract than in the detail.

In this case, compatibility is easier to achieve at the executive level than at the foreman level or, the more interfaces that are required, the greater the likelihood of incompatibility. Parenthetically, this suggestion is consistent with maintaining company identities as an objective in the negotiation process. This principle need not preclude informal technical contact at lower management levels providing that the contacts do not violate the terms of the contract(s). Similar arrangements have been shown to be practical in the interfacing of government and private organizations engaged in the same programs.[3]

A strange insight?

Considering the pressures and urgencies during merger negotiations, the next insight may seem strange to some. But it is a fundamental technique of architects and managers the world over that before beginning any program or closing any merger, they:

Pause and reflect
or, as bridge players would say,
"Review the bidding, please, before the next hand."

They know that answers will be needed soon enough to some highly charged questions. Of immediate interest to many stakeholders in the venture: should the present product lines and management architectures be continued, perhaps in parallel with the presumed ones for a while? What previously planned steps should be delayed, omitted, hastened, or just affirmed? Which units are most and least likely to experience significant change? More than anything else, questions like these should highlight the risk of a too-quick misinterpretation of the significance of the causitive event, itself. Still more quite general insights can help there:

> *The first quick look analyses and conclusions are often wrong.*
> *So, don't assume that the original statement ... etc.*

Conceivably the wisest move will, in retrospect, turn out to be an immediate and positive response to an offer to merge, but only if the prerequisites for decision making in the evolutionary past have been assured, if the foreseeable consequences are reasonably understood by all stakeholders, and if the chance to preempt the competition and to be the first success is really there. Unfortunately, such preconditions coming together at the same time are rare. An introspective pause just at this time may be just what is needed to avoid those infamous "serious mistakes in the first day." Remedying initial mistakes can be costly if not caught early. Waiting too long after a key decision can preclude using the *choose, watch, choose* insight (Chapter 7, pages 7-11) to reverse it. As Bob Spinrad warned:

> *Remember, that if you make (architecting) mistakes at this time,*
> *you will have to live with them for 20 to 30 years!* [SP 92]

Implementation

It is now time to actually make the change. Value judgments have been made by top management. Sufficient architecting has been done to declare the change to be worthwhile and feasible. The questions of why, when, and what have been answered. Now the question is: how?

A job for the professionals

First, a definition:

> Professional: One possessing great skill
> or experience in a field or activity.
> [Webster II 1984, 939]

In this case, professionals include specialists, managers, architects, engineers, and executives who qualify technically and/or administratively at whatever level is appropriate for the position.

If a radical change in course is to be successful then it is the professionals who must make it happen, given a clear statement of corporate purpose and status and a consensus of stakeholders that change is imminent and inevitable. Upper management can lead and architects can be instrumental in creating the vision, but if the center of gravity of human expertise of the organization doesn't move as needed, neither will the desired change.

The principal challenges to all professionals, each in their area of excellence, will be to think in new ways and to suggest solutions to new problems; in other words, to *innovate*. The architects will have to reinvent and improve the product line and organization, the specialists and engineers to create new client value, and the managers to form new teams and recommend the recruiting or phasing out of core expertise. Top management can listen and adjust at this point but arbitrary directives seldom can create widely accepted, practical solutions in time. Directing innovation externally has almost never worked. The reasons are basic. It is the individual dedication and voluntarily given time and energy that will move the center of gravity of human expertise in new directions, not external kicks and jolts.

It is the individuals who must trust and have confidence in their associates and clients with whom they must agree to work if the company is to succeed in its mission. In the upcoming knowledge era, the professionals as an interacting group will become the most productive wealth producers and the core capability of the company. In a very real sense, they will *own* the company because they will own its major asset — new knowledge and how to use it.[4]

Fostering innovation on demand during and after a radical change

What are the techniques of fostering innovation on call? First, it helps if there is a friendly willingness by one's superiors to consider and respond to new ideas, treating them as valued efforts to improve the organization, not as challenges to its management. If the ideas cannot or should not be implemented, it helps to explain why. More acceptable ideas may then be forthcoming. Were the earlier ideas inconsistent with company purpose or merely premature?[5] No matter which answer applies, the positive result is increased understanding and trust.

Second, allow and even encourage professionals to think about and educate themselves in fields related to those of the organization. If the organization is to move in new directions, the most likely ones are toward those fields that are nearby. In times of rapid change, it can be helpful to have at least a few professionals that understand "the fundamentals" of those fields. In that way, stumbles and gaffes will be reduced and entry into the field can be less amateurish and less subject to immediate rejection by the insiders, suppliers, and contacts on whom the organization may soon depend.

Third, begin serious exploration of other fields that can significantly affect present products — fields such as software, materials, systems, and lean production. Some burgeoning fields will prove easy to exploit and support: software-intensive vehicular transportation, Internet publishing and distribution, energy management, hard goods manufacture. Others may be wisps of somebody's imagination, likely to be unprofitable for the indefinite future: battery-powered bicycles, satellites generating electricity for use on the surface of the Earth, supersonic passenger aircraft for other than transoceanic flight.

And fourth, encourage management, particularly middle management, to think "outside the box" of their present positions; that is, to learn more about the organization's other units, imperatives, and problems. Many of the most productive innovations have come from simply establishing new relationships between units, the name of the game in all systems disciplines. If this outside-the-box thinking becomes normal and natural, it will surface very rapidly when the problems of abrupt external change calls for it. Older, once-premature, ideas may be quickly revitalized just when needed most.

Parenthetically, damaging professional loyalty and dedication by down-sizing, cost cutting, and the like, during or in conjunction with event shock, is as threatening to survival as an eagle breaking a wing just before the salmon season. Both eagle and a demoralized organization are then vulnerable to external change at the worst possible time. Needless to say, such damage strongly determines the future of even an excellent organization. If it loses its innovative drive, it loses its heart.

Management and policy resistance to change

Unfortunately, Hewlett-Packard's long history of innovation on need and call is rare. A more common story, dating back decades, is the DeSoto story. [RE 97 54] In brief, a young engineer was told by his boss not to propose innovative quality processes because to do so would violate the company's policy of planned obsolescence. The cars would last too long.

The problem it epitomizes, management resistance to innovation, is covered in some detail in Chapter 9, page 169 and Donnellon 94 and won't be repeated here. It has led to two especially relevant insights:

> *For every potentially successful innovation, competition will arise*
> *from those parts of the organization where there is perception of a*
> *resulting threat to the established system of organizational power.*
> And its extension,
> *The probability of implementing a new idea depends on the number*
> *of persons in the chain of approval.* [MI 92 27]

particularly in a presently successful operation!

Resistance to innovation, of itself, is not necessarily damaging. For potentially worthwhile innovations, and under the right circumstances, the effect of resistance may only be delay. Eventually, computers did indeed become effective, despite many misgivings about their error rate. Eventually, most homes had both electricity and telephones, despite the high initial investment and low quality. Eventually, environmental restraints were implemented over the real and imagined concerns of developers. But "eventually" can be a dangerous response once change becomes both inevitable and immediate. If a prompt response is an available option for a competitior, "too little and too late" may be the fate of the more resistant.

Merger successes and failures

From an architectural perspective, the intended objective of any merger is to add value to that of the companies considered separately; that is, added value or "synergy" will emerge from the combination. For example, a synergistic merger would add value if the business acumen of one company were added to the technical competence of another to produce an entity in which both parts benefited. An acquisition of one company by another solely to acquire certain skilled people, with no net value added, would not be considered synergistic.

One example of a synergistic merger was the successful one between the technology-oriented Collins Radio Company and the finance skills of Rockwell International, from which Bob Cattoi learned the following lesson:

> *When integrating two organizations, distinguish between*
> *the* real *synergies and the* perceived *ones early,*
> *and promptly activate the teams that can demonstrate that reality.*

Another example is the one between the Hughes Communications Satellite division and a company specializing in communication Earth stations, providing a "one stop" solution for customers who needed a complete space communication system; for example, the less-developed countries of the world.

Counter examples are those between General Motors and Ross Perot's EDS and, later on, between General Motors and Hughes Aircraft in both of which real synergy, if any, was far smaller than that predicted. The first merger collapsed in acrimony. In the second, the culture and forefront position in technology that Hughes represented were lost as General Motors placed profit first, as it had to, in an acquired company that had operated before the merger like a not-for-profit.[6] Neither stated rationale came to fruition, Hughes helping GM make better automobiles or Hughes operating autonomously with GM financial backing.

At least part of the reason was the sharp cultural differences between the companies involved. Whether that was recognized at the time by General Motors and Hughes Aircraft is not known to this author, but it was by both Rockwell International and Collins Radio and subsequently between Rockwell and Allen Bradley. Cattoi's lessons learned from all these mergers once again emphasized the importance of professional staff in any abrupt change of direction:

> *When integrating one structure or company into the structure*
> *of another, managing the soft assets of both*
> *is just as important as managing the hard ones.*
> and
> *Intellectual assets are the key to smooth transitions.*

Cattoi adds one more important lesson learned, one that echoes the "anchor" insight of pages 118 and 203:

> *Preserve the tradename and trademarks that are the rallying flag*
> *internally and the key hooks externally.* (Bob Cattoi 1998)

In actuality, the Rockwell mergers were successful for many other reasons, an important one being the decision by the Rockwell executive, Donald Beal (responsible for the mergers), that the cultures of the various Rockwell International acquisitions would be maintained, a policy which did much to retain and encourage the specialized professionals throughout the company. In the Collins case, the main interface change was at the finance level. Rockwell provided a larger and more experienced finance staff — a previous Collins weakness — resulting in a greatly improved core capability in that area for Collins and a better integrated Rockwell Collins enterprise to the benefit of both. The technical side of Collins, its primary core capability, was unchanged.

These decisions also minimized the temptation to combine operations whenever the two organizations' separate product lines happened to be essentially the same. One could say that these eagles didn't attempt to swim, they combined to catch more fish in a different way, each doing what each did best and which also benefited the whole enterprise as well. A true system and true synergy.

Final caution

Now a normal text would end here as the excellent and successful organizations and their people walked off into the sunset. But this author has been through six wars, six organizations, and at least as many "morning afters," some painful, to stop this text without a final caution.

> *It is not enough to win a war.*
> *You must also win the peace that follows.*

So, one last time into the breach.

Winning the peace by retaining the best

It is quite possible to accomplish a radical transformation in the short run that looks like a winner and then lose the desired results if the principal assets — the trust and confidence of the stakeholders in each other and in their professional expertise — are lost in the process.

The measurable "hard" assets of facilities, financial status, and annual sales can, of course, be treated as analyzable quantities and are merged, managed, and electronically traded as such. People, their dedications and their loyalties, as has been noted before, are not so easily traded. Unless considerable care is taken during the trauma of merging, some of the best and most mobile individuals leave, others consider leaving, and others worry about being forced to leave. Very few swear unswerving allegiance to the company in its period of uncertainty. It is indeed a sensitive and potentially dangerous time for the company as a whole and for the professsionals in it.

Counseling the pros

The author has many times — as a boss, executive, professor, relative, and friend — been asked for advice by highly qualified individuals anxious over their future. Some wish for greater compensation, others for recognition, still others for personal self-esteem and what they call "a reason for living," or a cause to follow. From advising these individuals, the author has gained a few insights that seem to work well. Yes, the author has tested them on his own quandries. As a general rule, the most anxious but risk-averse individuals decided to stay on their present course but said they now better understood why they were doing so. The most ambitious and risk-taking ones decided to change but only if their conditions were satisfied. Meanwhile these individuals quietly and purposefully developed the options to be able to change. As a CEO, these results were well worth the effort. As a professor, the effects were better students and more mature reports. As a friend and relative, I seem to have done no harm.

The insights by no means are new. The first one goes back at least to the early Greeks:

Know thyself.
Know what you like to do. Know what you can do well.

The second one is more action-oriented:

Keep trying to position yourself to be doing both what
you like *and what you* can *do well.*
If you do, you will "succeed."
So, don't worry about it!

It has been surprising to this author how many problems these two insights by themselves can resolve. The first calls for subjective introspection. The second calls for objective external assessment of what you do well, an invaluable calibration on one's feelings of self-worth.

A special group, those in a mid-life crisis, then ask, "What should I do next? You say I should do what I like and do it well. OK, I'll try. Success will come my way. Fine, but how do I leave if those criteria can't be met?" The two suggestions that I have given this group for years and which I learned the hard way are

Leave only when you know for certain why you are leaving.
Don't leave until you know where you are going.

The first is essential in avoiding the inevitable self-doubt of "what if..." The second demands a personal set of objectives and a strong desire to succeed. It also requires patience until a suitable position is offered. All these insights seem to apply whether the individual is thinking about leaving the organization entirely or simply about asking for a new task assignment.

What does this have to do with what excellent companies can and cannot do well? Isn't the success of the company dependent on the perceived success of its members? Isn't the feeling of personal success one of the best motivations for staying and for dedicating one's life and energy to whatever course the organization must follow? And, therefore, isn't it both an objective and a responsibility of the organization, in its own best interest, to be sure these personal concerns are acknowledged? For executives. For managers. For engineers and architects and specialists, for everyone. To manage task assignments, responsibilities, compensation, awards, facilities, personal mentoring, and education, fairness and respect is not just common sense, it can mean survival and excellence in times of radical change.

The catch? The organization needs to be prepared at least a decade before the trauma of radical change hits the compay and its people. May the reader never have to say, "If I had only known that 10 years ago."

Summary

Radical change is the ultimate test of an excellent organization. An architect would ask: does it know where it is and where it wants to go before the event shock that causes the inevitable change? Has it prepared for that change? Does it know which of its present strengths can be weaknesses in its future? Is it willing to change its culture and beliefs to achieve the change? Does it have the willingly offered expertise and dedication of its professionals that it will take to make radical change a reality? Is it prepared for radical change? If so, then get to work with clients, architects, and customers to create the product lines, processes, and supporting organizational structures that can make even a radical, unpredicted change a successful one.

Notes

1. May 14, 1986, May 28, 1986, and June 25, 1986 issues of *Orbiter* "Looking Ahead" articles published just after the Challenger accident and slightly more than a year before the author retired as CEO.
2. The close philosophical connections between organizations, their "languages," and software would predict that Gacek's results for software architecture are likely to be comparably true for organizations. [GAC 98]
3. The successful interface between the U.S. Air Force and The Aerospace Corporation in the area of human resources, one which requires aerospace people

to assist their Air Force counterparts in making technical decisions, reaches to the task level but not to the more detailed level of specifying the individual assigned to carry it out. The Aerospace Corporation is responsible for naming the individual and for being responsible for the quality of the work but not for deciding what should be done.

4. Note that this idea of professional ownership is quite different in the knowledge era from the worker-as-prime idea of the eighteenth and nineteenth centuries. Although workers created much wealth, they could not create land, take factories with them, or move from their communities at will. Their muscle, not their knowledge, was all they could bring to the bargaining table.

5. The author at one time was an aggressive advocate of a particular management change, thoroughly convinced of importance and need. The wise response by top management was that the idea was indeed excellent but, logically, it had to follow other changes in order to be effective. The author was convinced, existing measures were used to accomplish most of the objectives in the interim, and the originally proposed change, appropriately modified, quietly and without opposition, slipped into place within the year.

6. All profits, if any, were sent to a sole stockholder, the Hughes Medical Foundation, which generally reinvested the dividends in the company. No dividends were distributed either to executives or employees. While the company paid taxes on its earnings, for operating purposes it was managed and operated much like any not-for-profit organization.

Appendices

A. Propositions for organizational architecting

B. A listing of insights

C. Annotated list of citations

Appendix A

Propositions for organizational architecting

Those who would ignore history are doomed to repeat it.
[with a nod to George Santayana, 1863-1952,
who saw many changes in his lifetime.]

No man is an island and no organization is a world unto its own.
[with a nod to John Donne, 1573-1631, who lived on the
"tight little island" and should know.]

Introduction

About a quarter of a century ago, Frederick P. Brooks, now at the University of North Carolina at Chapel Hill, wrote what is now a classic work called *The Mythical Man-Month* [BR 75], *Essays on Software Engineering* from his IBM 360 operating system experience. With the encouragement of readers of that book, Brooks wrote a twentieth-year anniversary edition in which he extracted what he considered were the main points of each chapter, calling them, "Propositions of *The Mythical Man-Month, True or False?*" asking the reader to consider them when building software.

This appendix is written in that tradition. It is intended for those architects and clients, who are architecting an organization and need some suggested principles to follow, on how to proceed. Propositions are essentially the author's summary of key points of the text. Numbers following the propositions indicate the chapters where more information about each can be found.

There is necessarily some overlap with Appendix B's insights (domain descriptions and problem prescriptions). Though both propositions and insights are abstractions or distillations of materal found in the text, insights are lessons learned by many people and which have been "certified" by the criteria stated in the Preface and Introduction.

One might say that propositions are for architect–managers and insights for architect–professionals. An individual assembling a personal tool kit may well choose some of each.

The propositions

Group I: Architecting for constant change

- *Never* assume that the original statement of the problem is necessarily the best, or even the right one. It is only the beginning of a long discussion between client and architect. **(Chapters 5, 6, 9, and 10)**
- Choose the architecting team with great care, keep it small, and charter it well. It faces formidable obstacles. **(Chapters 5, 6, and 9)**
- Use insights to help make decisions in situations that are neither measurable nor replicable, otherwise use the well-developed tools of the applied sciences. **(Preface and Chapter 1)**
- Excellence and success are not the same. **(Part I Introduction, Chapter 5)**
- Understand excellence as dependent upon context, not as matching a general standard. **(Part I Introduction)** Being excellent in one field does not automatically mean being excellent in another. **(Preface)**
- Organizations can be better understood if viewed from more than one perspective; that is, as complex systems, as creators of emergent values, as competitors, as partners with government, as sets of beliefs, as structures, and as sets of interlocking decisions. **(Preface and Chapter 1)**
- Never forget that in complex organizations, virtually everything is connected to everything else, directly or indirectly. **(Chapter 1)** They have to be if emergent properties are to come forth from them all. Hence, a change in one almost always has consequences, often positive, in all the others. Be sure they are understood before acting. **(Chapters 2 and 10)**
- No organization can survive without a viable purpose. By the same reasoning, no viable purpose can survive without an organization designed to support it. They must fit each other. **(Chapters 3, 6, and 10)**
- Aggregating without adding value serves little purpose. **(Chapter 2)**
- Added value cannot last in the long run unless it is shared with not only the clients but with those suppliers that helped make it possible. For clients, it means better products at less cost. For suppliers, it means investment in a joint future. **(Chapter 3)**
- Pay special attention to the interfacing of different architectures, different software languages, and different organizational forms. In the interfaces are both the greatest added values and the greatest risks. **(Chapters 8 and 11)**
- Manage the architecting process carefully. **(Chapters 9 and 11)**

- Keep some anchors for people to hang on to during periods of change. (**Chapter 6**)
- Watch the order in which decisions are made; they can seriously affect the outcome.
- Listen carefully to all your stakeholders and keep them well informed. You will need their support and advice at critical times.
- Keep hardware, software, and process in synch. (**Chapter 11**)

• In an era of constant change, design evolutionary strategies, not only of product lines but of the processes and organizations that support them. (**Chapters 10 and 11**)

- Static, quarterly return-oriented strategies, created for static (commodity) products, can not be expected to match evolving product lines. (**Chapters 10 and 11**)

• Unless constrained, change has a natural tendency to proceed unchecked until it results in a substantial transformation of the system. In other words, it may take as much energy to end a trend as to start it. (**Chapter 6**)

Group II: Architecting for radical change

• The most disruptive changes are architectural or paradigm shifts in purpose, function, and form; between nonprofit and profit; between rule-based hierarchy and a free-form, "handshake" organizational structure; between public/government and private/business sectors; between high tech and commodity; or between product-orientation and process, profit, orientation, and cultural/service. Avoid such radical changes if at all possible. If not possible, try to change through architected intermediate states, reversible decisions, and escape options kept open as long as possible. (**Chapter 6**)

• Be prepared. Prediction may be impossible but preparation is not. (**Chapters 10 and 11**)

• Be prepared for strong resistance to change. (**Chapters 3, 6, 9, page 169 and 11, page 207**)

• When contemplating change, ask why, when, what, and how. Know beforehand your essential strengths and inherent weaknesses. (**Chapters 10 and 11**)

• To be successful, change, even radical change, takes time — a decade of preparation, a year or two to stabilize, and another decade to be firmly on a course. It will take the development of new product lines, new organizational structures to support them, new clients, new markets, new financing, some new staff and management, a recognized series of successes, and a re-earned reputation for excellence. The only real unknown is the exact time and circumstances. (**Chapters 1, 2, 3, 6, 9, and 11**)

- To be successful in radical change will require a diversity of perspective, experience, education, and beliefs. (**Chapter 6**)
- Be prepared to be inspired and to be able to call for innovation when needed. Foster innovation as a natural and accepted part of the organizational culture. (**Chapter 11**)
- Learn and use the language, rules, contexts, and imperatives of clients, competitors, and specialists, both present and potential. Make sure that your metaphors educate but don't alienate them! (**Part II**)
- Change takes time which can only be bought before it is needed.
  - Make the necessary advance payments including basic research, development of promising research results, education of the staff, architecting new product lines, software literacy, recruiting for the future, key personnel retention, "language" understanding, sharing added value, 10-year plans, and, last but not least, building long-term customer relationships. (**Chapters 1, 2, Part II Introduction, and 11**)
- Build a first-rate capability for software development and application both technically and managerially. It is the capability most likely to be needed in the twenty-first century of highly profitable, information-intensive, systems, processes, and organizations. (**Chapter 2**)
- No change can be called successful if it costs the dedication and inspiration of the primary asset of an organization, its professionals. (**Chapters 2 and 11**)

Appendix B

A listing of insights

Experience is the hardest kind of teacher.
It gives you the test first and the lesson afterward.
[Susan Ruth 1993]

Introduction: creating and using this list

From past experience with books on systems architecting, the favorite part of the books has often been the insight (or heuristic) list in the first appendix — in its way, a kind of "answers in the back of the book." They not only provide a quick summary of the main thoughts, they let the readers quickly find the context, references, and applications of each insight. These factors are essential if an insight is to be relied upon in use and not perceived as just another unsupported assertion. All insights (heuristics) in this text were subjected to the following criteria:

- An insight must make sense in its original domain or context.
- The general sense, if not the specific words, of an insight should apply beyond the original context.
- The opposite statement of an insight should be foolish, clearly not "common sense." That is, the opposite should not be equally sensible, nor dependent only on a special set of circumstances. Thus, neither "Look before you leap" nor "He who hesitates is lost" (while being chased to the edge of a cliff) would be accepted as "insights" in this listing.
- The lesson of an insight, though not necessarily its most recent for-mulation, should have stood the test of time and earned a broad consensus.

About 50 of the 126 insights in this book were taken verbatim from the earlier texts as being just as applicable to organizational architecting as to strictly technical product and process architecting. These insights, asterisked (*, #), may be found in the same form in [RE 91 312-316] and/or [R&M 97 232-243], respectively. The former, about 30, are with the permission of Prentice-Hall, Upper Saddle River, NJ. The latter, about 20 of which were not carried forth from the former book, are with the permission of CRC Press, Boca Raton, FL. The remainder, about 75, are either original in this text or separately referenced, or in such common use as to be unattributable.

The insights are organized under nine headings corresponding to the interests of organizational architects. They are purpose, success/failure, competition, constraints/rules, design, innovating, interfacing, test/diagnosis, and redesign/reorganization.

To speed lookup, two or three key words in each insight are set in **bold** type. Each insight is followed by the number(s) of the page(s), where it can be found in the text. The broadest or most general version of an insight begins in the left hand margin. Other, more domain-specific, versions are inset under the more general one. A few that are more appropriate to the broader propositions of Appendix A are subsumed under the appropriate proposition.

This listing of insights is only a part of all those that can apply to organizational architecting. Judging from past experience, as told in the introduction to Part IV, a few years of teaching them should yield hundreds more. Discovering, creating, organizing, and using a personal tool kit is left to the reader. Instructions on how to do so may be found in [RE 97 25-29]. In the meantime, please consider this list only as a starter kit.

Insights list

Purpose

- **Don't assume** that the original statement of the problem is necessarily the best, or even the right one. 97, 152, 169, 185, 205 * #
- The **purpose** of an organization and each of its organizational levels is to create capabilities and **add values** in addition to those of their separate elements. 23
- Unless you know **where** you are, you can't know where you are **going.** 191
  - If you don't know **where** you are going, you won't **get there.** 191
  - You must know where you've **been to know where** you're going. 153
- **Choose** as best you can. **Watch** to see whether solutions show up faster than problems. If so, the choice was probably a good one. But, if problems are showing up faster than solutions, revisit the decision that caused this to happen, and **choose** again. 134 * #

---

* [RE 91 312-316] With permission of Prentice-Hall, Upper Saddle River, NJ.
# [R&M 97 232-243] With permission of CRC Press LLC, Boca Raton, FL.

- The choice between the possibilities may well depend upon **which set of *drawbacks*** you can best handle. 164 * #
- No system can survive without a **viable purpose**. 23, 186
  - The test for economic viability: **Who benefits?** Who pays? Who supplies? Who loses? 61, 188 * #
- The most dangerous assumptions are the unstated ones. 95 #

Organizational and professional success and failure

- **Nothing fails like success.** 91
- Both **success and failure** are in the eyes of the **beholder** — and there are many beholders. 54 * #
- The test of a **good architecture** is that it will last. The sound architecture is an **enduring pattern**. 23, 103, 189 #
- The **bitterness** of poor performance **remains** long after the sweetness of low prices and prompt delivery are forgotten. 189 #
- **Great expectations**, because they are unlikely to be fulfilled, **define failure**. 101
- **Conservative expectations**, because they are likely to be accomplished, **define success**. 101
- **Value**, or worth, is in the eyes of the **beholder**. 15, 53 * #
- There's nothing like being the **first success**. 176, 187 * #
- The **pioneers** get the **arrows**. 176 #
- **Success is meaningless** if failure is impossible. (Andrew Jackson) 176
- The Law of **Unexpected Consequences**. 16
- Today, **wealth** is no longer best defined as ownership of land, goods, capital, or labor. It is (new) **knowledge** and knowing how to use it. 32, 34 #
- **Know thyself.** Know what you like to do. Know what you can do well. Then, keep trying to position yourself to be doing both what you like and what you can do well. If you do, you will "succeed." So don't worry about it! 210
- **Don't leave until** you know where you are going — it might not be there. Know **why** you are **leaving**, so that you never look back and ask, "What if?" 133, 184, 210
- **Experience** is the hardest kind of teacher. It gives you the test first and the lesson afterward. *x*, 219
- **Experience** is knowing a lot of things you shouldn't do. *x*
- To be successful requires a **diversity of perspective**, experience, education, and belief. 117

Competition

- **Timing. Timing. Timing.** 187

- It only takes one to **declare a race**. 56
- For every competitive system there will be at least one **countersystem**. 49, 53, 56, 98 *
- The efficient **competitor** looks for **mismatches** and tries to exploit or eliminate them. 53
- **In open competition,** the incumbent has the **encumbrances**. 38, 98
- Expect the **unexpected**. 16, 19 *
- A **system optimized** for a particular situation is **unlikely to encounter** it. 56 *
- It is **less painful** to anticipate than react. 152
- Sometimes the best way to defeat a threat is to do so **"out of bounds."** 57

Constraints and rules

- Given an excellent organization successful in its own field with objectives, skills, and policies designed for that success, there are some things it can*not* **do** — or at least not do well. *v* #
- The more rules there are, the **tighter** the constraints, and vice versa. 72
- **Unless constrained**, change has a natural tendency to proceed unchecked until it results in a substantial transformation of the system. 118 #
- **Bad cases** make **bad law**. 72
- A system (problem, idea) at rest will tend to stay at rest. **[inertia]** 55
- A system (problem, idea) in motion will tend to stay in motion. **[momentum]** 55
- To every action there is an equal and opposite counter-reaction **[force and energy]** 55
- For every management move there will be at least one countermove in reaction. 55
- Chances for recovery from a single management failure or flaw, even with complex consequences, are fairly good. Recovery from two or more *independent but coincident* failures is unlikely in real time and uncertain in any case. 189
  - **One failure** (or omission) acting alone can usually be corrected in real time. The system (organization) will probably survive. 9
  - **Two, acting together,** can usually be corrected, but not in real time. 9
  - **Three, because of their nonlinear interactions,** may never be corrected. The system (or the pilot) will crash before the cause can be determined. 9
- **Respect the rules**, the rulemaker and the rulekeeper, particularly in a period of dramatic change. 75
- Don't trip the balance of a good rule. It may fall over on you! 74

- **Forman's fact of political life.** 79 #
  - **Politics, not technology,** sets the limits of what technology is allowed to achieve. 79 #
  - **Cost rules.** 79 #
  - A strong, coherent **constituency** is essential. 79 #
  - **Technical** problems become **political** problems. 79 #
  - The best **engineering** solutions are **not necessarily** the best political solutions. 79 #

Innovation

- The three phases of innovation are **product** orientation, **profit** orientation, and **process** orientation. 200
- Inspiration is a function of preparation. Preparation is a function of purpose. Thus, **prepare to be inspired.** 202 *
- There is no such thing as **immaculate** communication! So, don't ever stop **talking** about the system. 200 *
- **Listen,** and remember that **communication** by definition is **bi-directional.** 200
- For every potentially successful **innovation,** competition will arise from those parts of the organization where there is perception of a resulting **threat** to the established system of organizational power. 207
- The probability of **implementing** a new idea depends on the number of persons in the **chain** of approval — particularly in a presently successful operation! 207
- Do the **hard parts first.** 48

Design of organizations and product lines

- **Don't assume** that the original statement of the problem is necessarily the best or even the right one. 152, 169, 185, 205 * #
- The architecture of the product, its construction process, and the **organization that manages them** should "fit" each other. 159
- All the **serious mistakes** are made in the first day. 49, 93 * #
  - The beginning is the most important part of the work. (**Plato,** fourth century B.C.) 93 #
- A picture is worth a **thousand words.** An insight is worth a **thousand analyses.** *ix, x,* 96, 151 #
- In architecting, the **order of decisions** can change the architecture. 130 #
- "I hate it!" is direction. Don't buck it or even try to. Learn from it. If the client reverses course, don't make a big deal of it. 46, 126, 164
- **Simplify.** Simplify. Simplify. 45, 152

- **Pause** and **reflect,** or, as **bridge players** would say, "Review the bidding please." 204 * #
- As for performance, cost, or schedule, **pick two** and reality will determine the third. 163
- Be prepared to **throw** the **first** (software) one **away.** You will anyway. 165
- Remember, that if you make [architecting] **mistakes** at this time, you will have to live with them for **20 to 30 years.** 205

Interfacing, integrating, and merging

- *No man is an island.* 45
- In complex organizations, almost **everything** is connected to **everything else.** 13
- The **efficient architect,** using contextual sense, continually looks for the likely misfits and redesigns the architecture so as to eliminate or minimize them. 48, 53, 147 * #
- The greatest **leverage, risks, dangers and opportunities** in a complex system or organization is in its interfaces and interrelationships. 166
- Choose boundaries with **low external complexity** (coupling) and high internal complexity (cohesion). 166 * #
  - In **partitioning,** do not slice through regions where high rates of information exchange are required. 9–10 * #
  - Design the sections, divisions, and departments to make their peformance as **insensitive as possible** to unknown or uncontrolable external influences as practical. 167
  - Choose boundaries with **minimal communications** required across them. 97, 120, 167
- **Relationships** among the elements are what give the organization its **added value.** 166
- Keep the **interfaces** simple and **unambiguous.** 45
- **Compatibility** is easier to achieve **in the abstract** than in the detail. 204
- Try to determine which **interfaces** are most **error-prone.** 167 *
- Communications suffer a loss of a **factor of 10** for every management level it transits. 35
- The **time** it takes to reach a final decision **doubles** with each **approval** required. 35
- When **integrating** one structure or company into the structure of another, managing the **soft** assets of both is just as important as managing the **hard** ones. 208

- When **integrating** two organizations, distinguish between the real **synergies** and the perceived ones **early,** and promptly activate the teams that can demonstrate that reality. 208
- "It's a **beautiful** thing when it's all working **together."** 32

Test and diagnosis

- **Murphy's Law.** If it can fail, it will. 27, 152 * #
- If it *can* fail, then fix it first! 27, 152
- **Deming's** advice: tally the defects, analyze them, trace them to the source, make corrections, keep a record of what happens afterwards. To which can be added: and keep repeating the process. 27 * #
- Toyota's **Five Why's.** 27 #
- **Everyone** a customer, everyone a supplier. 27 #
- Very **simple natural phenomena** can create what appears to be hopelessly complex (nonlinear) organizational behavior and confusion, in which constants aren't and variables don't. 17 (#)
- The first **quick look analyses** are often wrong. 95, 205 * #
- An insight is worth a **thousand analyses.** *ix,* 96 #
- For a system to meet its **acceptance criteria** to the satisfaction of all parties, it must be architected, designed and **built** to do so — **no more and no less.** 102 *
- Qualifications and acceptance tests must be both **definitive** and **passable.** 102 * #
- Before-the-fact **architecting** is much to be **preferred** over after-the-fact **assessments.** 189
    - Unless there is an effective **internal communication** system so that everyone who needs to know does know, somebody, somewhere, will **foul up.** 189
    - Unless everyone who **needs to know** does know, somebody, some where will **foul up.** 189, 200 *
- In **correcting** organizational problems, it is important that all the participants know not only **what** happened and **how** it happened, but **why,** as well. 189

Redesigning, rearchitecting, reorganizing, and merging

- **Reorganization** cannot be avoided, it is a natural part of **growth.** 159
- The **choice** between the possibilities may well depend upon which set of **drawbacks** you can handle best. 164 * #
- In times of change, **constants aren't** and variables don't. 17 #

- **Beware of evolutions** disguised as **revolutions**. More often than not, a new concept is another component in an old tool box, not another tool box. [Gold, Jeff 91] 187
- The exceptional **team** that created and built a presently successful product is usually the best one for its **evolution** — but seldom for **creating its replacement**. 98 #
- In introducing technological and social change, *how* you do it is often **more important** than *what* you do. 165, 197 * #
  - The **certainty of misery** is better than the misery of uncertainty. 197
- When implementing a change, keep some elements constant to provide an **anchor** point for people to cling to. 118, 203 #
  - Preserve the tradename and trademarks that are the **rallying flag** internally and the **key hooks** externally. [Bob Cattoi, Rockwell Collins 1998] 209
  - **Watch your image!** 115
- Given a change, if the anticipated actions don't occur, then there is probably an **invisible barrier** to be identified and overcome. 118 #
- A large change is best made through a series of smaller, planned, stable **intermediate states**. 120
- An element good enough in small system *or organization* is **unlikely to be good enough** in a more complex one. 6 * #
- It can be just as dangerous to **change too soon** as to change too late. 177
- The most successful firms both manage **incremental** change and are able to implement **strategic** change **prior to** experiencing (in anticipation of) performance **declines**. 189
- It's not enough to win a **war**. You must win the **peace** that follows (by retaining the best). 209

Appendix C

Annotated list of citations

Introduction

The following is an annotated list of books and articles cited in the text. It is intended to help the reader read further about specific subjects, rather than to be a general reference bibliography. The entries are annotated, alphabetized, and short-formed as [AB 97 22-29] in which AB are the first two letters of the author's name, 97 is the date in the twentieth century, and 22-29, if shown, are the appropriate page numbers. "USC Research Reports" are all by professional engineering students from the author's Systems Architecting course. They are unpublished but available from the University of Southern California, University Park, Los Angeles, CA 90089 — mail code 1450, Attn. Director, Systems Architecture and Engineering MS Program.

Listing

AB 97   Abd-Allah, Ahmed, *Composing Heterogeneous Software Architectures*. Dissertation Presented to the Faculty of the Graduate School, University of Southern California, in partial fulfillment of the requirements for the degree of Doctor of Philosophy (Computer Science) 1997. The initial work on which Cristina Gacek considerably expanded concerning the inherent incompatibilities between software architectures.

AD 97   Adams, Scott, *The Dilbert Future, Thriving on Stupidity in the 21st Century*. Harper/Collins Publishers, 1997. A satire on planning for the future.

AL 64   Alexander, Christopher, *Notes on the Synthesis of Form*. Cambridge, MA: Harvard University Press 1964. A classic treatise by a (civil) architect with a concentration on interface fits and partitioning.

BE 89   Beam, Walter H., *Command, Control and Communications* and "Evolutionary Systems: Arguments for Altered Processes and Practices." Alexandria, VA, The Beam Group 1989-90. Beam's perspective of architecture is as a continuing series of stable systems, roughly approximating a product line; that is, that architecting, development, and production have occurred successfully at least once, particularly appropriate for very large communications networks in which legacy is a strong driver.

BE 98   Bearden, David A., *A Methodology for Technicology-Insertion Analysis Balancing Benefit, Cost and Risk.* Presented to the faculty of the Graduate School, University of Southern California in partial fulfillment of the requirements for the degree of Doctor of Philosophy (Computer Science). The Darwinian evolutionary process is described by what is called a "genetic algorithm" which mimics birth, reproduction, and random cross-mixing of characteristics through mating, genetic mistakes, and survivability until a stable and satisfactory result evolves for the stated cost-estimation objective.

BER 97   Bernstein, Larry, *The View from the Cutting Edge or 319 Things I Learned in the Great Software Wars.* (unpublished essay) lbernstein@worldnet.att.net. 66 pages. A fascinating collection of one-liners from Mark Twain and Andrew Jackson to Dwight Eisenhower and of insights on software development. Bernstein is a retired chief technical officer of AT&T Network Systems. Available from author.

BR 75   Brooks, Frederick P., Jr. *The Mythical Man-Month, Essays on Software Engineering,* Anniversary ed., New York, Addison Wesley, 1975. Original edition of BR 95.

BR 95   Brooks, Frederick P., Jr., *The Mythical Man-Month, Essays on Software Engineering,* Anniversary ed., New York, Addison Wesley, 1995. The classic work on software architectures and engineering at IBM mid-1960s. Esp., the propositions, 229-250.

CA 90   Carpenter, Rob, *Biological architectures and organizational design for high-tech firms,* USC systems architecting research report. Los Angeles, CA, USC (unpublished). Points out how lessons learned in human growth parallel those in rapidly changing organizations, e.g., high-tech firms.

CH 98   Chappell, Sidney, et al., "New American manufacturers," *Auto. News Spec. Rep.,* April 27, 1998. An article on Japanese "transplants" into the "green fields" of the U.S. and their effect on the U.S. automotive market.

CO 98   Couretas, John, "Clueless in Seattle," *Auto. News,* October 12, 1998, 36-38. A description of the culture clash between Microsoft and the American automotive manufacturers.

CU 91   Cureton, Kenneth, "Metaheuristics," USC Research Report, December 9, 1991. The definitive work on metaheuristics — heuristics about heuristics — about 30 originals, including humanistic considerations, recognizing useful heuristics, selecting appropriate heuristics, applying heuristics, and reflection on results. Required reading for those constructing new heuristics and tool kits.

CU 92   Cureton, Kenneth, "The Behavior Patterns of System Architects (and Architect Teams)," USC Research Report, 1992. One of the best practical discourses on the use of the Myers-Briggs Temperment Indicators (MB-TI) for systems and organizational architecting. By a North American Rockwell manager-architect and adjunct professsor at USC.

DA 94   Davidson, William H., *Managing the Transformation Process: Planning for a Perilous Journey.* CEO Publication G 94-8, Los Angeles, CA: University of Southern California, 1994.

DO 94   Donnellon, Anne and Joshua Margolis, "The Delicate Art of Designing Interdisciplinary Teams," *Des. Manage. J.,* 5:3, Summer 1994, 3-14. The results of a survey of 12 teams in five *Fortune 500* companies with a summary that "interdisciplinary teams are delicate constructions subject to greater destructive force from external factors than from their more obvious internal disciplinary differences."

DR 81   Drucker, Peter F., *Toward the Next Economics and other Essays*, New York, Harper & Rowe, 1981. One of Drucker's best books on economics, business, innovation, and productivity.

DR 87   Drucker, Peter F., "The Creation of Wealth and What that Means to Organizations, Especially to Businesses," *Lord Foundation Award Winner Address*, Claremont, CA, Second National Symposium on Technology and Society, University of Southern California, Los Angeles, CA, February 1987, in which Drucker speaks to knowledge as wealth, curtly reviews the history of economics and economists, and recapitulates the Theory of Value in business terms.

FO 82a   Forrester, Jay, "A Technology Lag that May Stifle Growth," *Bus. Week,* October 11, 1982. Prescient article based on the author's expertise in modeling businesses.

FO 82b   Forrester, Jay, "A Black Cloud over the Smokestack Industries," *Bus. Week,* October 18, 1982. One of earliest predictions of subsequent restructuring of industry.

FO 97   Forman, Brenda, "The political process and systems architecting," *The Art of Systems Engineering,* CRC Press, Boca Raton, FL, 1998, 199-209. The original.

FR 92   Frericks, Jeffrey E., *A General System Architecting Methodology (A relational based architecting checklist)*. USC Systems Architecting Research Report, April 30, 1992 (unpublished). An effort to add rigor to the field by structuring all the steps taken by a systems architect during the definition phase. Partitions system architecting into four primary and 35 secondary parts within a context defined by seven input factors and two outputs, the system architecture and system architecting products. The top-level diagram of the methodology contains eight multiply-connected steps. Each step is further diagrammed. Clearly, whatever systems architecting is, it is not a simple discipline.

FU 99   Fulmer, Melinda, "Secret to His Success? Timing," *Los Angeles Times: Business Section,* February 9, 1999, C 1, 11. A brief report on Lowe Enterprises and its founder.

GA 45   Gardner, Murray F. and John L. Barnes, *Transients in Linear Systems,* vol. 1, John Wiley & Sons, New York, 1945, 2nd ed.. A classic work on the use of LaPlace transforms for linear systems.

GAC 98   Gacek, Cristina, *Detecting Architectural Mismatches During Systems Composition*. USC Doctoral Dissertation (Computer Science), December 1998. See also AB 97. Demonstrates 46 inherent incompatibilities between software languages (architectures) with the message that no two software architectures can be integrated because their constructs, assumptions, and languages simply don't match. Appears to this author (Rechtin) to forewarn organizational architects of similar incompatibilities especially in this era of software-connected organizational units.

GE 93   Geis, Norman P., "Profiles of Systems Architecting Teams," USC Research Report, 10 June 1993. One of the best researched papers on the composition of architecting teams by a Hughes Aircraft project manager–architect. Extensive interviews of top executives at Hughes, Raytheon,

Sparta, University of Pennsylvania, Martin Marietta-Orlando. Refers frequently to CU 92, its predecessor in this field of inquiry.

HAL 77   Hall, R. Cargill, *Lunar Impact, A History of Project Ranger*, National Aeronautics and Space Administration. NASA History Series SP 4210. Washington, D.C., 1977.

HAR 97   Harford, James, *Korolev, How One Man Masterminded the Soviet Drive to Beat America to the Moon*. New York, John Wiley & Sons, 1997. The story of the Soviet Union's chief space architect until his death in 1966. Shows how the Soviet Union lost the space race to the moon in the very beginning by management turmoil. See also OR 93.

HAW 83   Hawkin, Paul, *The Next Economy*. New York, Holt, Rinehart and Winston, 1983. Contains a good description of the long-period Kondratieff cycle and predicts the information age (the "informative economy") in some detail 10 years ahead of its more general concurrence. See also DR 81 and DR 87.

HAY 88   Hayes, Robert H., Steven C. Wheelwright, and Kim B. Clark, *Dynamic Manufacturing, Creating the Learning Organization*. The Free Press, New York, 1988, chap. 7. One of the few manufacturing texts that explictly addresses architecting of manufacturing plants, their software, and management.

HO 79   Hofstadter, Douglas R., *Gödel, Escher and Bach: An Eternal Golden Braid*. New York, Vintage Books 1979, 15-19. Gödel showed that no set of mathematical propositions can be complete in the sense of being provable from others in the same set. Although the book doesn't make a point of it, its underlying question is whether computers will ever think and, of course, what is meant by thinking.

JU 88   Juran, J. M., *Juran on Planning for Quality*. The Free Press, London, 1988. Defines design quality.

KI 91a   King, Douglas R., *Performance Models of Software-Intensive Systems: Their Falibility and Value*. Research Report #2 for ISE/AE 549A. Available through USC ISE Department, Dec. 1991.

KI 91b   King, Douglas R., *Software Life Cycle Process Primer*. Hughes Aircraft Company, Ground Systems Group, December 20, 1991. USC Research Report. One of the finest compilations of high-level heuristics for software cycle development, most of which are broadly applicable.

KI 92   King, Douglas R., *The role of informality in system development, or a critique of pure formalilty*. (unpublished) 1992 USC Research Report. The first to show the ties between philosophy and the art of systems architecting.

LA 87   Lang, Jon, *Creating Architectural History, The Role of the Behavioral Sciences in Environmental Design*, New York: van Nostrand Reinhold Company, 1987. A behavioral scientist's way of describing the modes within which a (civil) architect works. Although its examples are those of conventional architectural practice, the book's ideas are readily extended to systems architecting. Lang, at the time of the publishing, was a professor at the University of Pennsylvania.

LA 94   Lambert, Bob, "Beyond Information Engineering: The Art of Information Systems Architecture," Broughton Systems, Richmond, VA, August 22, 1994. Based in part on RE 91, it extends it to businesses. Identifies the essential skills of an architect as being a skilled business analyst, a designer of efficient feasible systems, a creative thinker, and a consensus builder. Indicates how to cultivate same.

LE 70   Leites, Nathan and Charles Wolf, Jr., *Rebellion and Authority, An Analytic Essay on Insurgent Conflicts*, Markham Publishing Company, Chicago, 1970. The classic on asymmetric warfare. Models both combatants and predicts results.

LO 89   Losk, Jonathan, "A Profile of the System Architect," (unpublished) USC Research Report, December 11, 1989. The first serious paper on the subject.

LO 90   Losk, Jonathan, "An approach for identifying potential system architects," (unpublished) USC Research Report. The starting point for research by Pieronek. [PI 90]

MA 88   Majchrzak, A., *The Human Side of Automation*, Jossey-Bass Publishers, San Francisco, 1988. Addresses the problem of the introduction and use of automation in the workplace.

MA 91   Madachy, Ray, *Formulating Systems Architecting Heuristics for Hypertext*, (unpublished) April 29, 1991. USC Research Report. A successful project to link all ER 91 heuristics using a Hypertext map (included). Results indicated feasibility up to 100 heuristics, including linkages across functional lines. Later superceded by "hardware store" metaphor for heuristics listing.

MAI 95   Maier, Mark W. "Heuristic Extrapolation in System Architecture," preprint. University of Alabama in Huntsville. ca 1995. Available from the author. One of the first articles to describe what became known as heuristic refinement from abstract to concrete.

MI 92   Mills, Archie W. Jr., "American Industry in the New World Order, Architecting for Innovation." 1992 USC Research Report. Begins with detailed stories of Grumann Aircraft, Lockheed Aircraft, Sony, and Honda and from them induces architectural insights on innovation. Well referenced, including books by Drucker [DR 85 and 86], Halliday, Morita, and others.

MO 93   Morris, C. R. and Ferguson, C. H., "How architecture wins technology wars, *Harv. Bus. Rev.*, March-April 1993. A seminal article which suggests a "Silicon Valley" model for victory and the use of licensing to expand a market.

NA 81   Nadler, Gerald, *The Planning and Design Approach*, New York, John Wiley & Sons, 1981, 3-7. One of the first purpose-first books on engineering planning and design. Introduces the idea of purpose-behind-the-purpose hierarchies.

NA 94   National Research Council, *On the responsibilities of architects and engineers and their clients in Federal facilities development*. Committee on architect–engineer responsibilities, Building Research Board, National Research Council, Washington, D.C. National Academy Press, 1994. Preface and Executive Summary. One of the most frank reviews of the painful area of government–supplier relationships. Should be required reading for all systems architects and their clients, even in other domains.

OB 83   Oberg, James E., *Red Star in Orbit, The Inside Story of Soviet Failures and Triumphs in Space*," A story told 20 years after the events with the partial release of information by the Soviet Union. See also HA 97.

OL 91   Olivieri, Jerry M., *Heuristics in the Development of Biomedical Devices*, (unpublished), December 1991 USC Research Report. The first of two reports on the differences between systems and biological heuristics.

OL 92   Olivieri, Jerry M., *Implementing Biological Models in Machines: Architectures, Applications, Heuristics, and Challenges*. USC Systems Architecting Report AE 549b, available from USC ISE library. A remarkable rethinking of systems architecting and its heuristics from a biological perspective. Rich in heuristics and their rationale.

PE 82   Peters, T. J. and Robert H. Waterman, Jr., *In Search of Excellence: Lessons from America's Best Run Computers*, Harper & Row, New York, 1982.

PH 89   Phadke, Madhav S., *Quality Engineering Using Robust Design*, Prentice-Hall, Upper Saddle River, NJ, 1989. An AT&T adaptation of Genichi Taguchi's methods for achieving quality at least cost.

PI 90   Pieronek, Tom, " A Personal Essay: The Search for Future Architects," (unpublished) USC Research Report. A survey of where in TRW most middle managers would expect to find individuals with systems architecting capabilities. Curiously, systems engineering was not first as a division, but fourth.

PO 95   Porter, T. M., *Trust in Numbers: The Pursuit of Objectivity in Science and Public Life*, Princeton University Press, 1995.

RA 51   Raymond, Arthur, *The Well Tempered Aircraft*, The Royal Aeronautical Society, London, 1951. Raymond is the source of the heuristic: *A good design is a compromise of the extremes.*

RE 91   Rechtin, E., *Systems Architecting, Creating and Building Complex Systems*, Prentice-Hall, Upper Saddle River, NJ, 1991. The principles of the systems architecting process, the challenges it addresses, and management issues it raises.

RE 92   Rechtin, E., "The art of architecting complex products," *IEEE Spect.*, October 1992, 29:10, 66-9. An article showing the close association of systems and architecting. Introduces heuristics to the general reader.

RE 94   Rechtin, E., Ed., *Collection of Student Heuristics in Systems Architecting, 1988-93*, University of Southern California, Los Angeles, March 15, 1994, (unpublished but available from USC on request for research purposes only). These insights, or systems architecting heuristics, were a primary result of professional engineer's reports on the lessons learned first from RE 91 laid up against their own projects and then from the lessons learned when students searched for additional ones that the projects themselves produced.

R&M 97   Rechtin, E. and Mark W. Maier, *The Art of Systems Architecting*, CRC Press, Boca Raton, FL, 1997. A focus on the problems addressed by systems architecting, the tools it uses, and its professionalization. Of particular interest is Chapter 8 on progressive design and progressive refinement of heuristics (by Maier).

RE 97   Rechtin, E., "The synthesis of complex systems," *IEEE Spect.*, July 1997, 34:7, 50-55. Argues that synthesis is very different from analysis. Introduces insights and metaphors to the general reader.

REN 95   Renton, M. B., *Organizations as Systems and Architectural Heuristics*, USC Research in Systems Architecting Report, 1995. A particularly insightful paper set in the context of planning the future of McDonnell Douglas

Aerospace. Renton was then a manager in the space transportation division.

RO 87   Rowe, Peter G., *Design Thinking*, Cambridge, MA, The MIT Press. Treats multiple and often dissimilar theoretical perspectives on how architects create and give shape to buildings and public spaces. Like Lang [LA 87], Rowe's ideas are readily transferred to systems architecting, though they arose from civil architecture. Considers diverse views from architects to theoreticians. Both books broke new ground in how architects and designers think. Rowe at the time of publishing was a professor at Harvard.

SH 96   Shaw, Mary and David Garlan, *Software Architecture, Perspectives on an Emerging Discipline*, Prentice-Hall, Upper Saddle River, NJ, 1996. A watershed text which helped lead software engineering into software systems (and systems software) architecting.

SI 69 & 81   Simon, Herbert, *The Sciences of the Artificial*, 2nd ed., Cambridge, MA, The MIT Press. 5th printing, paperback. A foundation of systems thinking, artificial intelligence, and the science of the mind in contrast with the "natural science" of how nature behaves. Page 132 notes the hope that a science of design can be created so that the field can attain institutional stature and credibililty, a hope this book's author sees as unlikely to be realized; much of architecting is not-science, not-mathematics.

SO 93   Sooter, Charles W., *Systems Architecting and Social Systems*. USC Reseach Report, 1993. A unique perspective on social system problems and the ability of systems architecting to help solve them.

SP 88   Spinrad, Robert, *Systems Architecture*, (unpublished charts for a USC lecture of November 1988). Available through its author, retired VP Technology Analysis Development, Xerox Corporation, 3333 Coyote Hill Road, Palo Alto, CA 94304. Succinct descriptions of systems architecture, its characteristics, and of major document processing architectural issues.

SP 92   Spinrad, Robert, *XEROX 2000*, (unpublished charts for a USC lecture of November 1992) indicating how Xerox was architecting its new course toward "The Document Company." Availability: see SP 88.

STE 83   Steiner, John E., "How decisions are made: major considerations for aircraft programs," *NAE Bridge*, Winter 1982-1983. Washington, D.C., National Academy of Sciences, 1983. A succinct history (1945-82) of decision making at the Boeing Company of which Steiner was vice president of corporate product development. Of general interest here are 7 of his 20 "Lessons Learned" (pages 32-3). In abbreviated form they are, using his numbering: (2) When the market is ready, the successful manufacturer may have to go. (3) A competitive loss sometimes motivates the loser more than it does the winner. (4) The highest validated [his emphasis] state-of-the-art has produced the longest term, and therefore, the most rewarding program. (8) The use of a parallel design team or "red" team to examine alternatives and to play a major devil's advocate role may be a valuable procedure. (12) Changes in a manufacturer's management team can change the company's responsiveness to problems and its ability to use company strengths. (16) Defining what constitutes "success" in the program is a necessary exercise. (17) A good airplane must be designed to meet a broad spectrum of market requirements. Compromises are essential.

TR 88   Trattner, Jon H., *The 1992 Prune Book, 50 Jobs That Can Change America*. The Council for Excellence in Government, 1998. The title word "Prune" is used to mean "Plum with Experience." A treatise on the qualifications of senior government officials.

TU 88   Tushman, Michael L., Wm. H. Newman, and David A. Nadler, "Executive Leadership and Organizational Evolution: Managing Incremental and Discontinuous Change," *Corporate Transformation; Revitalizing Organizations for a Competitive World*, San Francisco, Jossey-Bass Publishers, 102-130.

VE 93   Verma, Niraj, "Metaphor and analogy as elements of a theory of similarity for planning," *J. Plan. Ed. Res.*, 1993, 13:13-25. Directly addresses the question of the necessary conditions for credible metaphors. Argues that the metaphorical pairing of war and poverty (the War on Poverty) is flawed.

WE 84   Webster's II *New Riverside University Dictionary*, Boston, MA, The Riverside Publishing Company, 1984.

WH 89   Wheaton, Marilee J., "System architecting the software development process: the role of cost modeling," (unpublished) USC Research Report. The first of two on the relationship of systems architecting and cost modeling. December 1989.

WH 90   Wheaton, Marilee J., "Heuristics for Software Development and Cost Analysis," (unpublished) USC Research Report. Shows that systems architecting and cost modeling are closely related during conceptual design. May 1990.

WI 91   Witteried, Michael, *Application of Systems Architecting Heuristics to the Design of the DSCS Follow-On System*, November 4, 1991, USC Research Report. A study of a design competition for defense communication satellite systems.

WO 90   Womack, James P., Daniel T. Jones, and Daniel Roos, *The Machine that Changed the World, The Story of Lean Production*, Harper Perennial, 1990. An extensive study of modern, worldwide automotive manufacturing methods and the lean manufacturing methods that revolutionized them.

Glossary

This glossary is, in effect, a translation or reinterpretation in the language of the art of organizational architecting of the terms typically used in systems architecting and engineering. These fields, that of the art of organizational architecting, are sufficiently new that many terms have not yet been standardized. Usage is often different among different groups and in different contexts. The meanings below are as used in this book.

Abstraction

A selection of the presumably most important factors or features of a system or organization. Usually the features of organizations that particularly distinguish them from others.

Added value

The difference between the total value at the end of a process compared with the total value at its start. The difference between input value and output value of an organizational level or position.

ARPA (or DARPA)

(Defense) Advance Research Projects Agency

Aggregation

The gathering together of closely related organizational units, purposes, or functions.

Architect

A person doing architecting.

Architecting

The process of creating and building architectures, including those of organizations, especially during the conceptual and certification phases. Generally synthesis-based, insightful, and inductive. [RE 91 vi]

Architecture

The structure — in terms of organizational units, reporting and directing channels, factors or features, constraints and rationale — of an organization. Includes technical tasks, context, and outputs, as well as structure, people, strategy, processes, managerial, culture, and goals. See [REN 95 A10] and Part I, page 5.

Architecture, open

An architecture designed to facilitate addition, extension, or adaptation for use. In normal practice, the core of the architecture and its interfaces to others remains unchanged.

Architectural style

A form or pattern of design with a shared vocabulary of design idioms and rules for using them. Examples: hierarchy, matrix, segmented ring. See also RE 91 274-80.

CalTech

California Institute of Technology, Pasadena, CA

Certification

A formal, but not nessarily mathematical, statement that defines properties or requirements that an organization has met. Assurance by architect that organizational architecture is ready for use.

Chaos

Complex but structured behavior now known to result from nonlinearities, fixed time delays, memory, and interconnections. Found in such organizational contexts such as stock markets, inventory, and business cycles and communication networking. See also Part I, page 6.

CIA

Central Intelligence Agency

Client

The organization or individual that directly pays the bills; may or may not be the user.

Complexity

The degree of intricacy of a system so interconnected as to make analysis difficult or impractical. The more interconnections, the more complex the system.

Deductive reasoning

Proceeding from an established principle to its application. Characteristic of applied sciences and engineering and based on the principles of mathematics and science. Contrasts with inductive reasoning.

Design
The detailed formulation of the plans or instructions for creating an organizational unit.

Designing
One dictionary [WE 84 367] gives eight different definitions from conceiving in the mind to creating in a highly skilled manner. Designing generally means creating smaller objects in more detail with limited purposes, few interrelationships, and more artistic judgments than creating a large-scale system. See also Part I, page 5.

DOD (or DoD)
Department of Defense

Engineering
The process of applying science and mathematics to practical ends. [WE 84 433] See also Part I, page 5.

FCRC
Federal Contract Research Center funded by the Department of Defense.

FFRDC
Federally Funded Research and Development Center (all-agency term), see FCRC.

GM
General Motors

H-P (or HP)
Hewlett Packard Company, Palo Alto, CA

Inductive reasoning
Extrapolating the results of examples to a more general insight or principle. Characteristic of most of the arts and, in particular, the art of organizational architecting.

JIT
Just in Time, an inventory management technique.

JPL
Jet Propulsion Laboratory, part of CalTech under contract to NASA, 4800 Oak Grove Dr., Pasadena, CA 91109.

NASA
National Aeronautics and Space Administration

MBTI
Myer-Briggs Type Indicator. A way of characterizing the preferred ways different people have for solving problems. Related to, but not the same

as, so-called aptitude tests in acknowledging that preferences (what people like to do) are not the same as skills (what people are good at doing).

Megasystem or system of systems
A system composed of nearly autonomous, generally self-managed, subsystems each of which is capable of useful operation on its own, and which together produce results not obtainable from each separately.

Metaphor
A description of an object or organization using the terminology of another; for example, the use of a game (football, baseball, basketball, cricket, soccer) for describing how a unit or team can or should perform.

Model
A simplified but suitable abstraction of the key elements in an organization.

Nonlinearity
A property in which the result of introducing two inputs is not the same as the sum of the results of introducing them separately; in other words, they intermix.

Objectives
Client or organizatinal needs and goals, however stated.

Organization
Any organized group producing a product, process, or service including both private and public sectors, commercial, and military.

Organizational architecting and engineering
Systems architecting and engineering applied to organizations.

Paradigm
A scheme of things, a defining set of principles, a way of looking at an activity; for example, classical architecting methods, profit-driven management, cause-driven volunteer groups.

Perspective
A particular way of viewing a system, architecture, or organization, usually for a particular purpose such as physical structure, behavior, management, or cost.

Product
The result of an organization's activities for a client including manufactured goods, software, reports, technical support, etc. A single offering of a product line. Examples: the Ford 1995 Taurus.

Product line
The architected or designed framework on which a series of products are based. Usually an evolutionary, or open, design. Example: The Ford Taurus line from which comes 1995, 1996, etc. models.

Profit

An advantageous gain to the supplier.

Profit, for-

A description of an organization in which financial gains are returned for the most part to the investors, executives, and often employees. Also "bottom line."

Profit, not-for-

A description of an organization in which financial gains are returned to the organization and not to investors, executives, or employees. Examples: Rand Corporation, most universities, and colleges, The Aerospace Corporaton, charitable institutions.

Requirement

An absolute, nonnegotiable objective.

Risk

The perceived possibility of harm or loss. Generally refers to system- or enterprise-level losses of purpose or function.

Shuttle

A NASA space vehicle.

Specification

A statement describing what a device or system is to do, be, or meet.

Stakeholder

An individual or organization with a "stake" in the product, process, or organization. Not to be confused with "stockholder" whose primary interest is in lending capital.

Success

The delivery to a satisfied client of a promised set of results.

System

A collection of different things which, working together, produce a result not achievable by the things alone. See also Part I, page 4.

Systems (plural)

A term used to describe an hierarchy of elements each of which is not only a system but a subsystem to those below it and a subsystem to those above it.

System (or systems) approach

A management process in which virtually all decisions in all elements and subelements are made based upon the effects on the system and its functions *as a whole*.

Systems architecting
The art and science of creating and building complex systems, that part of systems development most concerned with scoping, structuring, and certification.

Systems engineering
A multidisciplinary engineering discipline in which decisions and designs are based on their quantitative effect on the system as a whole, based on applied science.

Technical decisions
Architectural decisions based on engineering feasibility.

USC
University of Southern California, University Park, Los Angeles 90089-1450 (School of Engineering)

Value
Worth. What you are prepared to give up to own. Can be financial, psychic, or both.

Value judgments
Architectural decisions based on worth to stakeholders, primarily clients.

Name index

Names in this index include not only those formally cited in Appendix C for directly related, documented contributions, but other individuals whose insights and ideas have influenced the writing of this book. Conversations with this author, seminal ideas expressed in a narrower context of a student report, and informal comments made in lectures often initiated trains of thought which never might have occurred otherwise. The page listings refer to pages in the text where the authors are mentioned either directly by name or indirectly through a specific citation in Appendix C. As examples, [MO 93] on page 64 refers to C. R. Morris and C. H. Ferguson by way of their 1993 contribution cited in Appendix C; (Fortuna 91) refers to an informal student report by David Fortuna in 1991. Generally speaking, the capital letters come from the first two in the individual's last name. The exceptions are second authors, which are listed under the first two letters of the first author's name. For example, Charles Wolf, Jr., a second author, is listed under LE (Leitis), the first author. Author references previously made in Appendix B are excluded in this listing.

The author of this book makes no claim that these sources first created these ideas — some clearly came from their own reading of others' works — only that they provide a path to the ideas for those interested in researching the field further.

To the extent that this Name Index suggests a conclusion, it is probably that the art of systems architecting emerged from the ideas of many people in many fields at many times.

Subject index

Introduction: Creating and using this list

This subject index is conventional with three exceptions. Not included here are names already listed in the Name Index, insights already listed in Appendix B, and citations listed in Appendix C. Using a single index which alphabetically lists everything from subjects, names, insights, and corporations has been found to be much less useful than presenting systems architecting from different perspectives; namely, of the basic ideas (the Table of Contents), of the tools (Appendix B), of the documentable sources (Appendix C), of who contributed ideas (the Name Index), and, finally, as subjects of interest to the field at large (this Index).

Corporations are listed only to the extent that their experiences provide insights of specific architecting interest. It would be a serious error for the reader to extend their experiences to evaluations of the corporations as a whole, either positively or negatively, if for no other reason than they have changed CEOs and managers many times in the past several decades. Without exception, all are regarded by this author as excellent — though not always successful in everything they try — or they wouldn't be in this book at all. At the same time, if a corporation is *not* listed, it does not imply either a compliment or criticism.

It is the nature of systems architecting that some subjects (ideas, concepts, corporations, systems, lean manufacturing, and "threads" such as quality and excellence) weave in and out of others and reappear in several different contexts. Indeed, this book could have been architected around them instead of the heading chosen. This Subject Index is intended for readers and specialists with such interests.

A

Acquisitions, 203
Advance Research Projects Agency (ARPA
    or DARPA) 66
  ARPANET, 67
  efforts to emulate, 81

Aerospace Corporation, 14, 96, 198, 203
  and the U.S. Air Force, 211
Apple Computer Company, 36
Architecting
  the art of, 140
  the basics, 145
  constraints on, 147